GLOBAL
TECHNOPOLITICS

The International
Politics of
Technology &
Resources

GLOBAL TECHNOPOLITICS
The International Politics of Technology & Resources

DENNIS PIRAGES

University of Maryland

Brooks/Cole Publishing Company
Pacific Grove, California

Brooks/Cole Publishing Company
A Division of Wadsworth, Inc.
© 1989 by Wadsworth, Inc., Belmont, California 94002.
All rights reserved. No part of this book may be reproduced,
stored in a retrieval system, or transcribed, in any form or
by any means—electronic, mechanical, photocopying, recording,
or otherwise—without the prior written permission of the
publisher, Brooks/Cole Publishing Company, Pacific Grove,
California 93950, a division of Wadsworth, Inc.

Printed in the United States of America
10 9 8 7 6 5 4 3 2 1

Library of Congress Cataloging in Publication Data
Pirages, Dennis.
   Global technopolitics: the international politics of technology
and resources/Dennis Pirages.
      p. cm.
   Includes index.
   ISBN 0-534-09912-2
   1. Technology—Social aspects.   I. Title.
T14.5.P57   1988
303.4′83—dc19   88-25887
                                                  CIP

Sponsoring Editor: *Cynthia C. Stormer*
Editorial Assistant: *Mary Ann Zuzow*
Production Editor: *Linda Loba*
Manuscript Editor: *Catherine Cambron*
Permissions Editor: *Carline Haga*
Design Director: *Katherine Minerva*
Interior Design: *Stephanie Workman*
Cover Design: *Sharon Kinghan*
Cover Illustration: *David Aguero*
Art Coordinator: *Sue C. Howard*
Typesetting: *Bookends Typesetting*
Printing and Binding: *Malloy Lithographing, Inc.*

Preface

The decade leading into the twenty-first century could well be described as the best and worst of times. For the minority of the world's population living in the industrialized countries, unprecedented levels of affluence have been attained and the liberal values guiding human behavior there embody some of the noblest aspirations of the human race. But for the majority of the world's population living in the less affluent parts of the world, exploding populations, limited economic opportunities, environmental deterioration, and decaying social orders make life for large numbers of people barely tolerable. And growing worldwide interdependence of many dimensions is linking the fortunes of people in these disparate parts of the world together in many unanticipated ways.

A number of tremors have shaken the increasingly interdependent international system shared by these diverse populations over the last fifteen years, raising urgent questions about their future prospects. Two oil crises, a raw materials and food crisis, and a bout of global inflation ushered in the period; a global debt crisis, stock market collapse, and serious deterioration of the earth's protective ozone layer ushered it out. The leadership position of this system's dominant actor, the United States, deteriorated steadily throughout the period. A relatively balanced U.S. trade situation declined into a series of staggering deficits—in excess of $150 billion annually—as the world's number one creditor nation quickly became the largest debtor. And the dollar, the de facto world currency, was severely battered by traders who seemed to have lost faith in the American future.

While the global political economy has seemed to lurch out of control, scholars and politicians have offered few satisfying explanations for the developing anomalies and discontinuities. Some have blamed big and inefficient governments for a seeming decline in growth and productivity, whereas others have found scapegoats in "unfair" trade practices of other countries. But comparatively little attention has been paid to ecological and technological changes as factors that have been quietly restructuring the international system in fundamental ways. While the ecological requirements of a wave of resource-intensive industrialization have created a series of global issues concerning resource interdependence, a new cresting wave of post-industrial technologies is raising sets of issues concerning structural and policy interdependence.

These overlapping waves of change, and the global issues they raise, demand anticipatory thinking by both scholars and politicians if the mounting problems of the twenty-first century are to be managed adequately. Existing institutions have a bias toward inaction built up over decades of relative prosperity and are ill-suited to deal with this acceleration of complexity. And academics, securely sequestered within the comfortable confines of established disciplines, currently lack the broad vision required to cope with the growing complexity they are unable to explain with the disciplinary tools on hand.

The chapters that follow are a modest attempt to develop an interdisciplinary theoretical framework for analyzing and anticipating changes in the international system and imposing some order on the growing tide of complex global issues. The first chapter lays out an anticipatory "inclusionist" framework for analyzing the interaction of techno-ecological factors and the international system. In the four following chapters, this framework is applied in turn to analysis of the likely impact of energy, food, and other natural resource discontinuities on the future international order. The last three chapters concentrate more directly on the impact of technological change on future relations among the affluent industrialized countries and on their collective relationship with that part of humanity currently experiencing the worst of times.

The author would like to thank Cindy Stormer, Linda Loba, Stephanie Workman, and others at Brooks/Cole for their help in producing this book. Thanks go also to the reviewers of the manuscript for their helpful comments: Professor Donald Baker, Long Island University; Professor Richard Bath, University of Texas; Professor Glenn Hayslett, University of North Carolina; Professor Sophie Peterson, West Virginia University; Professor Martin Rochester, University of Missouri; Professor Kenneth Rodman, New York University; Professor Catherine Wrenn, University of Colorado.

Dennis Pirages

Contents

7 The Politics of Technology Diffusion   168

8 Defending Future Generations   200

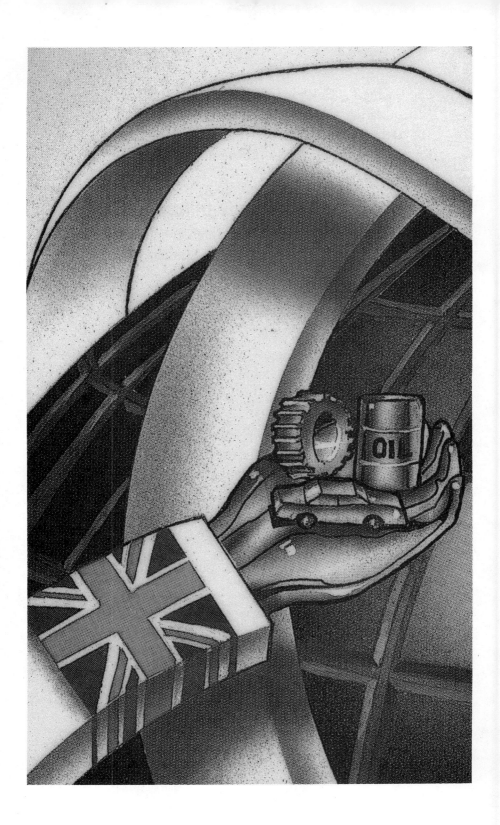

1

International
Relations
in Transition

The period of transition between the twentieth and twenty-first centuries is at once a fascinating and perilous time in international relations. It is characterized by large-scale technology-induced change, rapid population growth, and a related increased demand for natural resources to support industrialization over much of the planet. At the same time, the part of the world that has already reached an advanced stage of industrialization is undergoing a transition to some sort of post-industrial condition: population growth is slowing, service industries expanding, and natural resource-intensive production declining. These changes are causing significant social, economic, and political strains within nations and important transformations in relations among them. The next two decades may well mark a fundamental turning point in the global expansion of the industrial revolution as the human race grapples with two problems: how to deal with environmental limits to industrial growth on a global scale and how to manage new technologies that can yield a bountiful future for all or cause the instantaneous widespread destruction of civilization.

While a resource-intensive industrial revolution attempts to reach the far corners of the planet, a post-industrial revolution, driven by innovations in biology, telecommunications, and information processing, is reshaping international relations as well as the nature of the mature industrial countries. Internationally, nations once buffered from each other by oceans, mountains, and other natural obstacles are becoming part of an interdependent "global village" created by instantaneous communication, nearly immediate transportation, and integrated global markets. Events taking place in once remote corners of the world now often provide the main contents of the evening news in the industrial countries. The world's economies are presently so closely tied, by a highly integrated world trade system and associated capital markets, that domestic policies in the United States or any other major free market economy have significant consequences for countries around the world. In this increasingly complex and integrated global economy a sneeze in the United States, or Japan or West Germany, can cause much of the rest of the world to catch a cold.

A more ominous aspect of the march of new technologies is that new weapons technologies have given birth to highly accurate nuclear-tipped missiles, which now dot the world's landscape. These sophisticated missiles could reduce much of the world to rubble in a few minutes, but only limited political progress has been made toward eliminating them. In fact, weapons technocrats continue to attempt to peddle sophisticated "Star Wars" schemes, seeking the ultimate magic weapon that will obviate the need for negotiations over missile removal. In doing so, they threaten to ignite another round in a seemingly endless arms race and the spending contest that sustains it.

Because of the rapidity of contemporary technological change, and the magnitude of social challenges and potential problems it creates, there is now an unprecedented need for anticipatory thinking in both the public and private sectors. History continues to accelerate, and many decisions made today will be implemented in and have an impact on a much-changed world, which is now only dimly perceived.[1] Policy makers rarely take a long-term view of the consequences of their actions (or inaction) and are often caught unaware by unanticipated events that are part of this acceleration.

In the early 1980s in the United States, for example, economic factors and reactor incidents caused dozens of nuclear reactor projects under construction or on order to be terminated, creating an economic loss of tens of billions of dollars for the utilities involved. Many years earlier when decisions were made to build the reactors, few utility executives took the time to explore alternative visions of the future, to evaluate the technological problems involved, or to assess the nature of the economy a decade hence. Virtually none of them anticipated a world characterized by double-digit interest rates, an extended recession, falling petroleum prices, and lagging demand for electricity.

Internationally, few leaders in less developed countries (LDCs) applied a futures perspective to analysis of the course of world banking when, in the late 1970s, they began financing inflated petroleum bills with large, variable rate loans from multinational banks. Nor did the bankers, who were only too happy to lend those countries money without analyzing the long-term likelihood of default. The world debt crisis of the 1980s at least partly originated from this lack of foresight on both sides. And at present, many leaders in less developed countries, either from lack of foresight or for reasons of political expediency, seem to ignore the long-term devastating consequences of runaway population growth in their countries.

Intelligent management of change in the tumultuous, evolving international system requires an anticipatory perspective to understand the impact of contemporary decisions on very different future environments. In international relations, some understanding of the trajectory that the international system and its component parts are now following is essential in order to avoid "firefighting" as foreign policy and the related harsh consequences of failing to deal with present causes of future serious problems. Sobering future consequences will be paid if technological innovation is allowed to romp across the world's landscape unfettered by intelligent policies while problems of population growth, environmental pollution, resource depletion, and economic development grow more severe.[2] Unless contemporary leaders stop doing business as usual, the future world is likely to become "more crowded, more polluted, less stable ecologically and more vulnerable to disruption than the world we live in now."[3]

The collective ability of the human race to anticipate and cope with rapid change is limited. Prior to the twentieth century, the bulk of human experience took place during extended periods of institutional stability. Thus, the resulting social guidance mechanisms—the values, beliefs, and institutions that orient human behavior—are currently well suited to a stable and predictable world. But, because of a combination of rapid population growth, discontinuities in access to natural resources, new technological innovations, and related socioeconomic change, the context within which relations among people and nations take place now is rapidly departing from these historical standards; and people are bewildered when they find that following old prescriptions no longer leads to expected outcomes. It is therefore essential to develop foresight capabilities so that change may be anticipated and managed in the interest of future generations. Existing government institutions, being products of long periods in which change was not perceived as a problem, are ill equipped for an anticipatory mode of operation. Because of this lack of futures capability, it frequently appears that political leaders are napping in office—following rather than leading—and that events have somehow moved beyond their control.

A very basic reason for this dearth of anticipatory thinking is that the past successes of the industrial revolution and the peculiar history of abundance in the United States combined to create a pervasive antiplanning ideology. This ideology now covers over fundamental questions of purpose and organization and precludes many alternatives by claiming that laws of nature don't apply to human activities and that an undirected technological revolution will continue to provide benefits rather than create problems for future generations. Indeed, looking at past records of economic growth in the United States does yield a remarkable portrait of human accomplishment. Over the last two hundred years, a rapidly growing population has in general experienced substantial increases in life span and measured standards of living. And a sizable group of market-oriented technocratic optimists points to this record as an implicit guarantee that more of the same awaits the human race. Members of this caste often argue that anticipation and planning interfere with an "invisible hand" responsible for this remarkable growth and that if society strays from the "natural" course of events it will lead to trouble. In the typically optimistic words of Herman Kahn and Julian Simon,

> Global problems due to physical conditions . . . are always possible, but are likely to be less pressing in the future than in the past. Environmental, resource and population stresses are diminishing, and with the passage of time will have less influence than now on the quality of human life on the planet. . . . Because of increases in knowledge, the Earth's carrying capacity has been increasing through the decades and centuries and millenia to such an extent that the term carrying capacity has by now no useful meaning.[4]

If the less affluent parts of the world could only continue to follow this progress trajectory for the indefinite future, there would be little reason to engage in speculation about potential future problems. Humankind would continue to triumph over nature and new technological innovations would meet all human needs and wants. But there is a substantial contending body of evidence and opinion that sees the great tide of materialist industrial affluence coming to an end. Transformationists foresee a much different world shaping up in the twenty-first century, a new world that could mark a sharp discontinuity with the growing affluence of the industrial period. In the words of *The Global 2000 Report to the President,* published as a joint effort of several agencies of the U.S. government, "Serious stresses involving population, resources and the environment are clearly visible ahead. Despite greater material output, the world's people will be poorer in many ways than they are today."[5] Others, such as futurist Willis Harman, write more directly about various fundamental dilemmas that result from the problems of technological success and look forward to a major transformation in the accepted way of understanding contemporary world problems.[6]

It isn't necessary to reject all the arguments of the technocrats or accept the positions of the transformationists in order to appreciate the need for more and better anticipatory thinking. The bitter conflict between the two camps over alternative views of the global future is reason enough for clarifying studies, if nothing else. But this whole agenda of fundamental issues of growth, purpose, and progress has been and still is ignored because it isn't perceived as pressing in a political world of two-, four-, and six-year terms of office. The more pristine academic world also fails to respond to the challenge of anticipatory thinking, because it is divided into cohesive groupings called disciplines within which analysis proceeds from a segmented, technocratic point of view. The study of planetary futures is interdisciplinary by nature, but academics don't communicate well across disciplinary lines. Thus, time passes, global problems continue to mount, and little is done; policy makers react to each new crisis with a fire-fighting mentality, ignoring the old maxim that an ounce of prevention—anticipation?—is worth a pound of cure.[7]

▼ An Inclusionist Perspective

In the face of this acceleration of events and growing problems related to population, technology, and resources, the analysis of international relations needs a theoretical perspective that lends itself to interdisciplinary thinking and that offers the potential for anticipating complex future problems before they develop. This theoretical framework should be useful in synthesizing the perspectives of different disciplines toward the development of a more holistic theory of international behavior.

Human beings perceive and scientists and philosophers study the nature of the world and social relationships within it in three different domains. The most concrete and mutually verifiable observations are made of a physical domain, or nature, within which the evolution of Homo sapiens has taken place. In this biophysical domain, hereafter called the *techno-ecological,* scientists study the tangible world and humankind's place in it. Chemists, physicists, biologists, and ecologists, for example, aim to discover the physical laws of nature and the related biophysical evolution or natural selection that occurs in nature.

In another less visible domain, best called the *structural,* social and behavioral scientists study the interactions of human beings and the persisting behavior patterns that organize and guide human behavior. In this social realm, human beings share experiences and define social reality as individuals, groups, and nations.[8] It is in this domain that a process called social evolution, parallel to natural selection in the physical world, takes place.[9] Two of the main goals of social or structural inquiry are to better understand how different societies structure themselves in response to nature's challenges and technological change and to improve the human social condition through better understanding social and behavioral processes. Thus, in this domain economists study the exchange of goods and services among people and countries; political scientists examine power relationships, both domestic and international; sociologists look at various aspects of human social relationships; and so on.

In a third and still more abstract domain, best referred to as the *realm of the mind,* psychologists and philosophers labor over the least observable aspects of social life, studying belief systems, values, and human purpose. Workers in this domain analyze information processing, conceptual frameworks, and terminal values that guide human behavior. The aim of this type of inquiry is to gain an understanding of human motivation and to speculate about the evolution of survival-relevant cultural information, which provides guidance in coping with problems of change.

Scholars generally devote their efforts to increasing specialized knowledge in small corners of one of these three domains. In each of them, some communication among academics in the various disciplines takes place, and some hybrid fields, such as biochemistry or political economy, even develop. But communication among scholars in different domains is very rare. Thus, both human thinking and theories of human behavior tend to ignore causal relationships that may exist among the three domains. Disciplinary explanations or, at best, causal explanations restricted to a single domain are common. For example, philosophers speculate about the nature of human values, belief systems, or ideologies without relating their origins to conditions in the physical world; economists generally fail to acknowledge ecological factors that shape and limit economic performance; and political scientists often ignore the biological and ecological basis of individual and nation-state behavior.[10] Thus, in the study of international relations many scholars consider ideologies and belief systems immutable, see little linkage between ecological changes and various kinds of structural instability, and in general fail to understand the dynamics of system change.[11]

In the real world there is a very close linkage among the ecological, structural, and value domains, and the limited ability of the various disciplines to focus on these links is an impediment to increasing scientific understanding. The evolution of human societies and relationships among them has taken place within tight constraints posed by nature. Factors such as population changes, competition with other species, limited and shifting resource bases, and the inexorable march of technology have had a significant impact on human social arrangements. Nature and technology combine to provide basic sets of rules that Homo sapiens, like all species, is forced to respect.

Theories that attempt to explain interactions of human beings in nation-states are also best anchored in an understanding of the basic ecological principles governing relationships of all species to nature.[12] Figure 1-1 lays out some potential causal relationships among the three domains. At the base are techno-ecological variables, such as population, climate, natural resources, and technology. Throughout most of history, these have been nature's givens and beyond human control. The resource base found within national boundaries is a given. Oil fields and iron ore deposits are products of nature, not of human wishful thinking. Similarly, population growth has been for the most part beyond human determination until very recently, and it still is for much of the world's current population. Although technological innovation is, in theory, a factor that can be accelerated or depressed by human choices, the historical pace of technological change seems to have been governed as much by the rate at which nature has yielded secrets as by human endeavors. In this respect

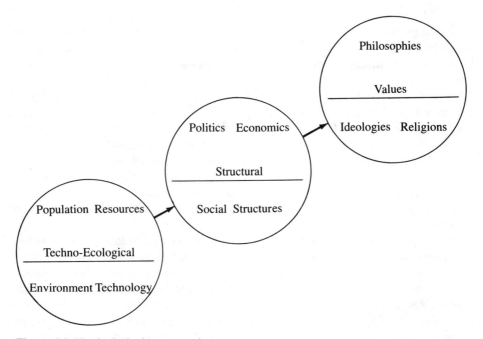

Figure 1-1 The inclusionist perspective.

science and technology can be considered to be embedded in nature. To say this is not to claim that population and technology have never been and cannot be subjected to human control and subordinated to human values, but only that prior to the contemporary period much of the time they were nature's givens.

The structural and value domains are important because they determine the perceived quality of human existence and also include values and institutions that routinize and organize human behavior. Human societies have developed a bewildering array of institutions and moral prescriptions for accomplishing similar tasks, and some of them have been much more effective than others. The values that guide human behavior differ from one society to the next, but some are more functional than others in orienting human behavior to the realities of the physical world.

Taken together, the structural and value levels of a society compose its culture. Cultures are the products of social evolutionary experiences, which operate in the structural and value realm as a sort of natural selection mechanism. A culture provides members of a society with a cognitive map, or an explanation of what the world is about and their places within it. It also has a normative component, explaining the difference between good and bad. Finally, a culture contains an evaluative component, a guide for everyday action to be followed by the good member of the society.[13] Cultures and related world views are continually being shaped and reshaped by technological and ecological changes; some cultures adapt to these changes much better than others.

The atypical four-hundred-year stretch of recent history dominated by the industrial revolution has given rise to a culture or predominant world view that could best be called *exclusionist*. This world view permeates social life and consciousness as well as academic enterprises in the industrial world. It is manifest in the beliefs that human beings exist apart from nature and that they are destined to dominate it, that the last four hundred years of growth and progress will be followed by a similar stretch of good fortune, and that there are few limits to resource-intensive industrial growth. This position, in a sense, exempts human beings from the laws of nature.[14] It is buttressed by a kind of arrogance conceived in materialist industrial successes, embraced by technocratic optimists and manifest in their beliefs that whatever human beings put their minds to can be accomplished. It has been a predominant view in recent presidential politics, the Reagan "Star Wars" campaign being but one example of attempts to use technology to solve problems that really require political and social solutions.

This exclusionist perspective has also conditioned the development of the social and behavioral sciences. The social sciences are rooted in the abundance and anthropocentric thinking that has accompanied the industrial revolution.[15] Social inquiry usually begins by assuming that human institutions and values evolve quite apart from any technological or ecological underpinnings and ignores the question of human adaptation to biophysical exigencies. This optimistic perspective in inquiry, which Dunlap calls the human exemptionalist perspective, now pervades the mainstream social science that has been flourishing during the advanced stages of the industrial revolution.[16]

An opposite, and much more humble, point of view is taken in this book: the human inclusionist perspective. This view focuses on humankind as one among many species competing for the earth's limited resources. It is oriented to analysis of change and sees human institutions and values as products of social evolution that has been heavily influenced by changes in nature and technology. This perspective is embraced by many transformationists, who argue that the period of the industrial revolution has been unique in human history and has very little future. In politics the inclusionist perspective was substantially embraced by the Carter administration in its *Global 2000* study as well as in its more general concern over the future of the global environment. The inclusionist perspective stresses the pressing nature and sociopolitical impacts of global issues such as rapid population growth in the face of finite supplies of natural resources. Inclusionists do not see an unregulated marketplace as an ideal way of meeting human needs, given the seriousness of environmental abuse, and they argue that economic activity must be managed carefully if destruction of nature is to be prevented. This nascent inclusionist world view is offered as a post-industrial competitor to the materialist exclusionist view that continues to shape the thinking of "industrial man."

The academic world has not been particularly responsive to the challenge of developing an inclusionist social and behavioral science. But a small group of scholars has sketched out the elements of a post-industrial paradigm for the social and behavioral sciences stressing the close ties between the physical world and social policy.[17]

The inclusionist strategy for the development of theory stresses the causal preeminence of techno-ecological variables and their impact on structures and values. Put simply, the causes of change are taken to be in ecological and technological variables and reflected in changes in structures and values. While the inclusionist point of view stresses causal relationships among the techno-ecological, structural, and value domains, it recognizes the stabilizing and guiding role played by existing values and institutions and thus generates theoretical propositions about change in probabilistic terms. It is useful as a framework for analysis of change that takes place in behavior, values, and institutions over long periods of time; it makes no pretense of explaining individual events in the short term.

Anthropologist Marvin Harris has analyzed the evolution of cultures from a similar, "cultural materialist" point of view. His world is also divided into three parts or domains—the infrastructure, structure, and superstructure—which are roughly akin to the three domains in Figure 1-1. In the Harris analysis, the infrastructure is made up of those things in nature that are largely beyond human control, such as climate or natural resources. The infrastructure represents "the principal interface between culture and nature, the boundary across which the ecological, chemical and physical restraints to which human action is subject interact with the principal sociocultural practices aimed at overcoming or modifying these constraints."[18] His structural domain is made up of routinized patterns of social, economic, and political behavior that serve as the communicators of social evolutionary experience. The superstructure refers to the symbolic processes that are important for the human psyche.[19]

Harris argues that social research strategies are best anchored in a methodology that gives strategic priority to infrastructure as shaping changes in structure and superstructure.

> The order of cultural materialist priorities from infrastructure to the remaining behavioral components and finally to the mental superstructure reflects the increasing remoteness of these components from the culture/nature interface. Since the aim of cultural materialism, in keeping with the orientation of science in general, is the discovery of the maximum amount of order in its field of inquiry, priority for theory-building logically settles upon those sectors under the greatest direct restraints from the givens of nature. To endow the mental superstructure with strategic priority, as the cultural idealists advocate, is a bad bet. Nature is indifferent to whether God is a loving father or a bloodthirsty cannibal. But nature is not indifferent to whether the fallow period in a swidden field is one year or ten.[20]

To put matters more succinctly, Harris makes a strong case for an inclusionist strategy in analyzing cultural change based on the constraints nature imposes on the forms that social structures and values may take. Ideas and institutions can proliferate in great number, but those that give improper guidance in dealing with technological and physical reality are likely to perish quickly.

The only difference between the Harris logic and that suggested in Figure 1-1 is in the role ascribed to technology, but this difference is more apparent than real.

Technology does play an important mediating role in altering the features of the physical environment and thus in modifying relationships between nature and society. But at any given point in history technological possibilities for changing nature's parameters are limited. Thus, scientific and technological discoveries can be conceived of being as much a part of nature revealing itself as products of human ingenuity.

The inclusionist perspective increases understanding of the interaction among the techno-ecological, structural, and value domains and offers a parsimonious strategy for anticipating and explaining change in human societies and relations among them. Just as the bodies of contemporary human beings represent the products of hundreds of thousands of years of biological evolution, social structures and collectively held values are to a great extent the products of a parallel social evolutionary process.[21] The human body has adapted to a changing physical environment through natural selection processes based on differential rates of reproduction. Those genotypes best adapted to the physical environments experienced have survived in greater numbers and thus determined the nature of succeeding phenotypes. Social evolution is likewise a dynamic process by which values and forms of social organization adapt to changing technological and physical realities.[22] The resulting culture—values, ideas, concepts, and information—can be looked upon as survival rules, or a guidance mechanism produced by evolutionary experience. Structures (institutions) transmit this experience across generations and are themselves products of the social evolutionary process.

This inclusionist strategy is not determinist and does not suggest that human values, thoughts, or motives be looked upon solely as products. Rather, it argues for a probabilistic strategy that gives priority in building theory to a causal flow that starts with techno-ecological factors, moves to the structural realm, and ultimately reaches human moral codes, beliefs, and values. By contrast, exclusionists would argue for reversing theoretical priorities. Relationships among the three realms in reality are very complex and there is a great deal of feedback among them. Furthermore, core human values—those Rokeach refers to as terminal—unquestionably persist over long periods of time and become important factors in shaping the structural and ecological realms in the short and medium terms.[23] But terminal values, like other aspects of the value realm, are over time likewise shaped by the social evolutionary process. In an unchanging world, structures and values would be completely congruent with pressures from nature and technological conditions and thus provide optimum guidance for preserving stability. But in the real world, the physical environment and technologies are always in flux, and congruence with structures and values is only imperfectly maintained through continuous adaptation.[24]

Finally, and most important, this inclusionist strategy can be directly applied to an anticipatory framework useful in analyzing global futures. A great deal of information about futures in the techno-ecological domain has been generated by a series of studies undertaken over the last decade. For example, reasonably accurate projections of world population growth are available by country and region for the next three decades. Barring some unforeseen global disaster, demographers confidently expect about 7.8 billion people to be living on the earth in the year 2020.[25] Although there is less certainty and agreement about the amount and location of the world's crucial natural resources, there is reasonably good information about energy reserves, water supplies, arable land, pollution levels, and so on.[26] Future technological developments are less understood and predictable, but the scientific foundations for several post-industrial technologies have been laid, and their impact on future societal and global value systems and institutions can be anticipated with reasonable certainty.

Far fewer futures studies have been done analyzing the international system in the structural and value domains, and correspondingly less is known about future changes in rules and values that will govern international behavior. Since futures can be and have been projected in the techno-ecological domain with reasonable precision, studying how these changes have impacted and will impact the structural and value domains can yield clearer perceptions of future problems and conflicts in the international system and its component parts. For example, the recent slowdown of population growth and the related "graying" of populations in the industrialized countries suggests the likelihood of less political and economic dynamism in those countries and possible evolution of more conservative values. Juxtaposed with rapid population increases and related youthful populations in the less developed world and the potential development of radical change-oriented values there, inclusionist logic leads to expectations that these divergent demographic trends would lead to significant clashes in values between North and South in the twenty-first century. Similarly, the lack of large petroleum and natural gas reserves in Western Europe in the face of abundant, but capital and technology-intensive, reserves in the Soviet Union would lead to predictions of pressures for greater future cooperation between those regions based on energy–technology swaps over the next few decades.

The inclusionist anticipatory framework, giving priority to techno-ecological factors in explaining change, is employed in the rest of this book. An attempt is made to synthesize what is known about likely future changes in the techno-ecological domain and to link those changes to changes in the international system, within the nations that make it up and in the rules that govern interactions among them. Thus, likely future patterns of population growth, resource availability, and technological innovation are examined, and their impact on the future international system and its component parts is anticipated in the sections that follow.

▼ Technology, Revolutions, and Social Paradigms

Many scholars interpret the last decades of the twentieth century as a departure from established patterns of social evolution and the beginning of a period of revolutionary change in mature industrial societies.[27] The term *revolution* has a myriad of meanings and is often used loosely. Here it refers to large-scale discontinuities in the structural and value realms of societies. There have been only two such major revolutionary transformations in world history that have impacted the majority of the earth's population. The first large-scale revolutionary transformation of human culture was the agricultural revolution, which apparently began simultaneously in several different places in the Middle East around 8000 B.C. and then subsequently spread slowly outward to the rest of the world. The second was the industrial revolution, which began to gather momentum in the fifteenth and sixteenth centuries in Western Europe and is currently spreading spasmodically to the more remote areas of the world.

Both of these previous revolutions were driven by significant technological innovations and related ecological changes, which created social surplus, or capital over and above what was needed for subsistence.[28] The agricultural revolution was initially driven by innovations in farming that permitted humans to exploit nature more efficiently. These innovations included domestication of plants and animals, and they provided social surplus resulting in enhanced diets, more dependable food supplies, and more sedentary populations. The slow spread of the agricultural way of life, with its related value system and world view, was accompanied by modest population growth, development of permanent settlements and small cities, and a growing division of labor that was instrumental in producing still more surplus. The agricultural revolution eventually produced enough surplus in some countries to support military castes and organized warfare and culminated in the development of several large empires.

The industrial revolution gained much of its impetus from technological innovations that utilized fossil fuels—coal, petroleum, and natural gas—to do the work previously done by human beings and draft animals. This second revolution, which now has moved into advanced stages in much of the world, originally produced social surplus of unprecedented magnitude by enhancing human productivity through a much more intricate division of labor feeding on a generous natural subsidy stored in coal, petroleum, and eventually natural gas.[29] New political, economic, and social institutions were shaped by these technological changes as significant social surplus permitted unprecedented social and political change. While the advanced stages of the industrial revolution have been accompanied by rising material expectations, production of goods has increased apace, giving rise to a liberal world view that seeks social stability through increased production rather than management of demand. On the negative side, however, much of the surplus thus generated has been consumed in two massive world wars and a subsequent arms race.

The tumultuous events of the last twenty years are evidence that resource dislocations are pushing and new technologies are pulling the world into a third revolutionary period. But unlike the previous two revolutions, which swept aside tradition on the strength of new technologies and related economic benefits, as yet no positive vision has been articulated by advocates of a new way of life. Thus, the industrial countries seem caught between two ages, experiencing the economic stagnation and political uncertainty brought on by the worldwide slowing of industrial growth while not having yet developed a positive vision and related values appropriate for a new age.

The mental product of these revolutions and related long periods of social evolution, generations of human learning experiences translated into social survival rules, can be called a dominant social paradigm (DSP).[30] These survival rules—individual beliefs, ideas, and values transferred from one generation to the next—are carried as part of the larger culture and transmitted through socialization processes. During normal times of slow change, only simple maintenance learning is required to sustain a dominant social paradigm. Indeed, in the absence of perceived threats to established ways of doing things, it is very difficult to avoid problems of social stagnation. The industrial paradigm or world view has been shaped by generations of material-intensive economic growth and seemingly gives technological optimists a recipe for future success. But just as natural selection in the biological realm cannot anticipate future environments—witness the fate of the unfortunate dinosaur—the cumulative survival information contained in the present dominant industrial paradigm may not give relevant guidance for coping with the turmoil of a post-industrial transformation.

The dominant paradigm concept was developed in the work of Thomas Kuhn, who applied the idea to the intellectual frameworks that hold together the various scientific disciplines.[31] He argued that normal disciplinary research is largely based on a scientific paradigm shared by scholars—a collective understanding of facts, rules of inquiry, standards for evidence, and problems in need of solution. Shared paradigms are important in the sciences because they give ready definition to the very complex world in which scientists operate. They provide a means of organizing research and passing on basic information from one generation to the next, thereby obviating the need for each to start from scratch.

Full-blown scientific revolutions, or paradigm shifts, are like social revolutions and occur very rarely, because vested interests—paradigm proponents—can be persistent even in the face of clear anomalies. One of the best examples of the trauma that can be involved in a paradigm shift in the sciences is the Copernican revolution in astronomy. Prior to Copernicus, astronomers invested a great deal of effort shoring up the orthodox Ptolemaic paradigm, a theoretical framework that placed the earth at the center of the universe. Astronomy became an increasingly precise science, and data started accumulating that could be fit into the Ptolemaic model only with the greatest of difficulty. Anomalies began to mount, yet those who had invested a great deal of work within the established paradigm went to often absurd lengths to incorporate the findings into their intellectual frameworks. Quite ludicrous and very complex models of the solar system were created in an attempt to shore up the old geocentric way of looking at the world. Eventually, anomalies became so obvious that a new view of the solar system, one that places the sun at the center, became accepted after much controversy as a new organizing paradigm for astronomers.[32]

A dominant paradigm in the sciences, according to Kuhn's argument, provides a framework within which scientists choose research topics, search for data, and interpret experimental results. A dominant paradigm can be thought of as a pair of glasses, or blinders, that clarifies, or obfuscates, a very fuzzy and complex empirical world. It defines the "is and ought" of the scientific profession, the problems worthy of solution and the methods by which they can be solved. Paradigms remain dominant in the sciences for long periods because the reward systems within disciplines encourage conformity with established scientific norms. A good scientist accepts the givens of the profession, doesn't ask unseemly questions or rock the boat, and is usually rewarded with cash, power, and scientific prestige.

There are significant parallels between scientific and social paradigms. The dominant social paradigm is the survival-relevant component of a culture: the collection of norms, beliefs, values, habits, and related rules of interpretation that provide a frame of reference or world view for members of a society. A DSP is a shorthand way of talking about the predominant world view, model, or frame of reference through which individuals and societies interpret the world around them. It defines the nature of the physical and social worlds as well as the "is and ought" of society, including rules for survival and related ethical principles.[33] Dominant social paradigms change slowly during normal times characterized by long periods of stability. But social paradigms have changed rapidly, or shifted, in response to techno-ecological stimuli during the worldwide spread of the two revolutions mentioned earlier. A DSP defines social reality and shapes social expectations. A paradigm shared by a majority is essential for social stability: without shared norms and values and a consensus view of the world, people would be constantly quarreling with each other over fundamental definitions of the good life.

Values and value systems, which provide guidance for appropriate action, are key components of social paradigms. Rokeach has defined the value realm as consisting of terminal, instrumental, and peripheral values. Terminal values are most critical to paradigm persistence, since they define preferred ends, or goals in life for individuals. Rokeach contends that there is "a relatively small number of core ideas or cognitions present in every society about desirable end states of existence and desirable modes of behavior instrumental to their attainment."[34] It is this core of values that guides behavior during normal times and rapidly changes during revolutions. A dominant social paradigm provides value continuity, although within any DSP small and adaptive shifts in values are constantly taking place. It is only when fundamental obstacles to adaptation—such as new technologies or changes in the physical environment—are encountered that paradigms and component value systems break down.

▼ Cracks in the Industrial Paradigm

Change in dominant social paradigms can be slow or rapid, destructive or constructive, minor or major, depending on the rate at which failures and anomalies mount up and the ability of leaders to respond with appropriate policies. Over the course of the industrial period, constant small changes have taken place within the overriding world view that now defines the industrial DSP. But in the last two decades of the twentieth century, considerable evidence is mounting to indicate that the paradigm currently prescribing and shaping behavior within and among nations is not providing adequate guidance for coping with rapidly changing technological and environmental realities.

The current cracks in the industrial paradigm are not easily noticed, because in the early stages of a paradigm shift it is difficult for people caught up in it to realize that a transition is taking place. But such periods are detectable, because they are times when old rules no longer seem to apply and conventional explanations for mounting anomalies no longer make sense. Under such conditions of uncertainty, individuals may return to old ideologies, authority figures, and religious practices in a desperate attempt to cope with an increasingly unexplainable and threatening world. While the speed of the current paradigm shift is not certain, there is ample evidence that the practices, norms, values, and theories that have been most associated with material-intensive industrial expansion are frequently being challenged.

Harman has suggested four key dilemmas or anomalies within the industrial paradigm that cannot be resolved without a major system transformation.[35] The *growth dilemma* has developed because economic expectations associated with the industrial paradigm require continued rapid growth in material consumption, while on a planetary scale it is impossible to live with the consequences. The *control dilemma* results from a need for more government capacity to control techno-ecological change at the same time that social values portray such control as evil and a brake on progress. The *distribution dilemma* revolves around a growing need for wealthy individuals and nations to develop mechanisms to help those less fortunate while in reality the rich attempt to insulate themselves from the rising expectations among the growing numbers of the world's poor by building walls around their spheres of prosperity. Finally, a *work roles dilemma* results from the fact that not enough industrial jobs can be created to keep unemployment low and meet the expectations of future generations. Harman links these four dilemmas to the emergence of a new scarcity. Old scarcities involved shortages of things needed to meet basic human needs and were overcome by developing new technologies and using more land and resources. The new shortages, however, result from ecological scarcity, approaching technological and resource limits to growth on a planetary scale.[36]

Manifestations of the politics of new scarcity and the approach of ecological limits have been obvious in international relations over the last two decades. The twin oil crises of 1973–74 and 1979–80 and subsequent events in the Persian Gulf have called attention to the longer-term problem of petroleum and natural gas depletion as well as the more immediate problem of reserve location. The food and basic commodity crises of the mid 1970s indicated how rapidly an overheated global economy can exhaust resource inventories and drive up prices. The food crises in Africa in the 1980s highlighted the anomaly of a world in which farmers in industrialized countries hold land out of production while millions starve for lack of food. The global recession, debt crisis, and stock market crash of the 1980s stressed the interdependence of capital markets and the need for careful management of surplus (capital) available for investment on a global scale. Even the growth dilemma has become more obvious in recent statistics. From 1960 to 1973, the real growth rate of the industrial economies was about 5 percent. Over the next decade this growth rate was cut in half.[37]

Aside from the push of natural resource shortages, the decline of old technologies and the pull of new ones are combining to create very different types of post-industrial societies as well as a new kind of international competition. Many observations have been made about the limited returns on additional investment in the resource-intensive technologies characteristic of the heyday of the industrial period. Renshaw, for example, has cogently argued that many of the processes responsible for industrial progress are subject to diminishing returns and can no longer be pushed to yield the same degree of economic growth. Speed in transportation has been pushed to the point where it is no longer economical to move any faster. Each additional mile per hour costs more in fuel expended than the amount gained in human productivity. Similar limits are found in various types of automated production and in the efficiency of increased scales of production and distribution.[38]

The overall rate of resource-intensive industrial innovation is slowing down, and innovations are largely restricted to the service sectors of the mature industrial economies, according to Giarini and Loubergé.[39] Others have made an interesting case that the period leading up to the twenty-first century might usher in a depression related to the declining phase of a Kondratief long wave of technological progress, which could result in a depression like that of the 1930s.[40] These and many related assessments of the acute problems of advanced industrial societies could well indicate that the industrial revolution has now peaked and that a "third wave," or high-tech revolution focusing on telecommunications, biotechnologies, and information processing, is now under way.[41]

▼ International Relations: The Great Transition

The industrial revolution that originated in Western Europe has been responsible for a huge surge in the human population, massive economic growth, and a rapid worldwide spread of Western influence. A three-hundred-year period beginning about 1650 A.D. produced a new international order predicated on rapid growth and the expansion of Western European economic and political dominance. During this time the population of the world increased eightfold, consumption of fossil fuels skyrocketed from nearly nothing to more than seven billion metric tons of coal equivalent annually, the quantity and variety of nonfuel minerals required dramatically increased, and contacts among various parts of the world greatly expanded.

The historical roots of this demographic, economic, and political expansion and the development of the accompanying industrial paradigm can be traced back to twelfth-and thirteenth-century Europe and an initial outward surge that accompanied the feeble beginnings of industrialization. Immanuel Wallerstein has identified two periods of outward thrust from Western Europe following on the heels of innovations in transportation and communications.[42] The first occurred between the eleventh and mid-thirteenth centuries and resulted in the recapture of Spain from the Moors, the conquest of Sardinia and Corsica by Christian zealots, the addition of other territories as a result of the Crusades, and an English expansion into Wales, Scotland, and Ireland. The second and better known outward thrust was the Atlantic expansion of the fifteenth and sixteenth centuries, which was motivated by the need for gold, spices, fuels, and food staples in Western European nations.[43]

According to Wallerstein, the pressure for the second outward expansion came mainly from a combination of scarcity and greed.[44] The fourteenth century was not a pleasant period for most of the European nobility, since populations were growing and the amount of cultivated land was not. Opportunities for growth in production within various kingdoms declined as marginal lands—those not really fit to yield a reliable harvest—were forced into production in an attempt to sustain incomes. There are also indications that climatic fluctuations added to agricultural production problems.[45] A seemingly endless series of small wars made increased taxation and government spending a necessity, which further cut into the prerogatives of the nobility. Given general declining growth prospects, an increase in conspicuous consumption among the nobility, and a need for more food and resources to meet the requirements of growing populations, a push into the less technologically sophisticated areas of the world was almost inevitable.

The Portuguese were the first actively to acquire colonies, a fact that can be explained by a dearth of domestic expansion opportunities. They were soon followed by the Dutch, British, Germans, French, and eventually other European countries in expanding influence into distant areas of the world. Many of the colonized Atlantic islands became sources of wood and sugar for European populations pressing close to the carrying capacity of their own lands. The American colonies provided lumber, silver, gold, and tobacco. The continuing wave of expansion eventually carried European armies into the most remote corners of Africa and Asia, where natural and human resources were seized in the name of kings, queens, and economic growth.

A second wave of European expansion in the fifteenth and sixteenth centuries resulted from more rapid population growth, additional technological innovation, rising economic expectations, and lack of domestic resources and opportunities.[46] During this latter expansion, a significant portion of Western European economic growth could be attributed to colonization of much of the non-European world. Of the more than 175 non-European nations and territories that exist at the present time, the vast majority were colonies only three or four decades ago. At the beginning of World War II, nearly one third of the world's land area and population was to be found in colonial possessions. European countries—Great Britain, the Netherlands, France, Belgium, Portugal, Italy, and Spain—with a combined population of about 200 million persons controlled over 700 million people in their empires. Japan controlled 60 million and the United States about 15 million.[47]

Choucri and North have used the term *lateral pressure* to refer to the growth dynamics that led to Western European expansion into foreign territories. As they put it, "When demands are unmet and existing capabilities are insufficient to satisfy them, new capabilities may have to be developed. . . . Moreover, if national capabilities cannot be attained at reasonable costs within national borders, they may be sought beyond."[48] Lateral pressure can be manifest in positive and negative forms. On the positive side, expansion of trade with other countries is an option for meeting domestic wants and needs without resorting to violence. Political confederation or development of common market arrangements is another. But during the second wave of expansion and during subsequent periods of history, nations possessing a technological edge often have used military force to build the colonial networks they have used to meet their needs. The route that countries have followed in getting resources beyond national boundaries has been a function of domestic needs, economic and military capabilities, comparative levels of technological sophistication, geographic location, and the power, friendliness, and resources of neighboring states.[49]

From an inclusionist perspective the environmental and technological pressures driving this second historical global expansion seem obvious, but the colonizers at the time were not social scientists and weren't particularly introspective about their motives. Some justified the lateral expansion in the name of religion: bringing Christianity to the "heathens" in Africa and Asia. Others found justification for these exploits in economic and commercial rhetoric. Still others rationalized these adventures as manifest destiny, an imperative to explore and conquer. Colonies became accepted as appendages of the colonizing countries, and colonialism as a natural state of affairs. At the time of this great Western expansion, there were few moral inhibitions about taking territories and resources from others. Power and might made right, and expansionary activities were halted only when the imperial ambitions of one country intersected with those of another.

The industrial revolution was thus accompanied by an expansionist international politics driven by techno-ecological imperatives. Over time, the dependence of colonizing countries on colonial networks continued to rise as populations and the demands of industrialization grew apace. The nationalist independence movement, which gathered momentum in the post–World War II period, came as a major political and economic shock to those countries that had come to count on the colonies as sources of cheap human and natural resources essential for further economic growth. But at the end of World War II the former colonial powers were drained of the capital and manpower required to maintain colonial order. In the face of this growing independence movement, the costs of maintaining order became prohibitive; sometimes peacefully, sometimes violently, the colonies slipped from bondage and an international system quite unlike the predecessor, with its Western European and U.S. dominance, began to form.

In summary, the industrial revolution created an international system dominated by the technologically advanced countries. Eventually a world system, including colonial empires dominated by Western Europe and Japan, evolved within this hierarchy of nations. At the peak of colonial expansion, nearly 90 percent of the non-Western world was controlled by the colonial powers.[50] The surplus generated by large-scale industrialization, however, not only enhanced living standards but was eventually used to finance two incredibly expensive world wars, the second of which triggered a dissolution of empires and a transformation of the world system. In the last decades of the twentieth century, as part of more general challenges to the industrial paradigm, the previously disadvantaged colonies are attacking the hierarchy, the purpose, and the rules of conduct of the dissolving international system of empires.

▼ Theoretical Paradigms in International Relations

Scholarly analysis of international relations has developed within and therefore is difficult to separate from the dominant social paradigm of industrial societies. International relations theory has been very much influenced by a Western industrial world view that is closely related to more general developments in the behavioral and social sciences. Theories, concepts, and research are dominated by an exclusionist view of the world that may well lend itself to inappropriate interpretations of current trends and events.

The history of international relations as a discipline, or subdiscipline, has been characterized by a search for a unifying scholarly paradigm. The most commonly accepted explanatory framework within the field is an exclusionist, state-centered, power-politics view of reality. It is firmly anchored in explanations originating in the structural and value domains and depicts nations as frequently in conflict because leaders cannot refrain from attempting to exercise their power within the international hierarchy. According to this view, these industrial-paradigm political "realists" aggressively pursue short-term national interests leading to inevitable conflicts among the nations making up the international system.

Mansbach and Vasquez have characterized the essentials of this exclusionist, state-centered paradigm as follows.

1. Nation-states and/or key decision makers and their motives are the starting point in accounting for international political behavior.
2. Political life is bifurcated into domestic and international spheres, each subject to its own characteristic traits and laws.
3. International relations can be most usefully analyzed as a struggle for power. This struggle constitutes the major issue occurring in a single system and entails a ceaseless and repetitive competition for the single stake of power. Understanding how and why that struggle occurs and suggesting ways for regulating it is the purpose of the discipline.[51]

Mansbach and Vasquez conclude their analysis of the traditional paradigm by observing that, in essence,

> in domestic politics a single actor, the government, has sufficient power to regulate
> the activities of all other entities in society, producing a certain measure of order
> and tranquility; in the international arena, there is no such leviathan. Consequently,
> in such an "anarchic" environment, each nation-state must struggle to maintain,
> if not to increase, its power; otherwise it will be crushed.[52]

Although this description may be somewhat overdrawn, these basic ideas have guided generations of international relations scholars and provided policy prescriptions or ideological justifications for political leaders as diverse as Joseph Stalin and Ronald Reagan. The inadequacies of this framework are found in its exclusionist nature, which stresses the autonomy of decision makers and limits causal explanations to leaders and their motives. It assumes that leaders operate in a vacuum rather than an environment of pressures and issues shaped by technological and ecological change. Its basic weaknesses have been illuminated, over time, by numerous scholars, who have contributed pieces of a possible new explanatory paradigm to replace the one now in a state of decay.[53]

The emphasis in this book is on building an inclusionist approach to the theory and practice of international relations that could provide the glue needed to hold many of these new ideas together. This approach attempts to link together the three domains described above and gives theoretical priority to explanations originating in the techno-ecological domain. Nations are viewed as organized human populations coping with physical laws and resource requirements similar to those confronting other species. Nations compete and cooperate within a system composed of a physical ecosphere, a structured set of practices and rules referred to as the international political economy, and an ideological realm characterized by value conflicts among proponents of different organizational methods. How well nations perform is primarily a function of the ways that they organize internally and the domestic and international policies that they choose to deal with nature's challenges and the imperatives of technology.

Global technopolitics, then, refers to the dynamics of an emerging post-industrial international system increasingly driven by the imperatives of technology. It is different from the expansionist system of the industrial period in a number of ways. There is now no country clearly in charge of maintaining order and prescribing rules of conduct within the international system. The former hegemony of the United States has been repeatedly challenged in different ways by Western Europe, Japan, the Soviet Union, newly industrializing countries such as Korea and Taiwan, and more recently by coalitions of less developed countries. The political economy of unlimited economic growth and expansion of Western influence has become one of limited opportunities, as population growth, unstable energy prices, capital shortages, and slower economic growth have sobered developmental optimism. This contemporary period of transition to a post-industrial international system is not unexpectedly beset with crises: a food and raw materials crisis, two energy crises, a protracted world debt crisis, and a collapse of world economic confidence. And given the onset of near military parity between the United States and the Soviet Union, new economic conflicts, rooted in the development and implementation of new technologies, have emerged as a source of friction in the industrialized world. Thus, the nature and exercise of power are changing dramatically as "low politics" economic, ethical, and ecological concerns are replacing "high politics" issues involving potential use of military force.

Technopolitical considerations, now growing steadily in importance in relations among nations, were previously neglected because leaders and nations in a position to set the international agenda thought it best to ignore them. One of the most important tools of power is the ability to set agendas and to keep certain issues from coming to public attention. It is a clear sign of the growing power of the former colonies that they have been able to shatter the old consensus and call attention to a series of previously neglected issues. This has been especially obvious at the United Nations, an institution that has been transformed from a forum dominated by the United States and Western Europe into an organization that now gives extensive exposure to Third World points of view.

The nature of power is also being redefined as part of the more general transformation taking place in relations among nations. In the old industrial hierarchy of nations, naked military force and the characteristics of a nation that supported the use of such force were considered the essence of power—a colliding-billiard-ball approach to power analysis. Hans Morgenthau, for example, became famous for his long list of factors that supposedly increase or decrease the ability of a nation to use force against others. Among his factors were geography, industrial capacity, population size, national morale, an adequate resource base, and the quality of government.[54] It is not that technopolitics makes military power unimportant. Rather, the international rules of good conduct have changed so that the naked exercise of force has become much more difficult for mature industrial countries. And the focus has shifted away from gain by conquest to gain through technological domination and resource manipulation.

The limits of power have also become more clear as its application is often now circumscribed by the tremendous power that the two superpowers possess. They are wary of getting into situations that could escalate into a mutually annihilating nuclear exchange. They also have learned, albeit very slowly, that in some situations, like Afghanistan and Vietnam, military technology cannot determine the results. It is a little ironic that the spread of nuclear weapons from the United States to the Soviet Union, Great Britain, France, China, and India has made these nations more cautious about getting involved in military adventures and has created significant freedom for smaller powers to maneuver beneath a resulting nuclear umbrella. Thus, although military expenditures continue to soar on a worldwide basis, the willingness of the major powers to use military power has diminished.

Another factor that has eroded the willingness of major powers to use military force is the high costs involved in an era of increasing economic austerity. Major conventional wars are very costly affairs. They require participants to have strong economies and access to cheap resources. Environmental contraints, high prices for fuels, limited supplies of capital, slow economic growth, and a host of other factors increasingly constrain major power options. The United States severely stretched economic limits to support the Reagan military buildup in the 1980s, and possibilities for having both guns and butter are increasingly circumscribed. Although the lesson has yet to be learned by some very senior political leaders, national security can be jeopardized as much by unmanageable economic and psychological drains caused by preparation for war as by military adventures.

Finally, perhaps the most important change in the perception of power is the increasing role played by nonmilitary technologies and control of natural resources. National security is increasingly defined as economic well-being, requiring steady economic growth, a competitive effort in civilian research and development, and access to required natural resources. In the simpler days of industrial expansion, a nation's power was military—one-dimensional and easy to understand. In the emerging interdependent world of the future, characterized by global technopolitics, power will be a much more multidimensional concept, and maintaining national security will be defined in much broader terms than simply maintaining sizable military forces.

▼ Notes

1. See Gerard Piehl, *The Acceleration of History* (New York: Knopf, 1972).
2. See Donella Meadows et al., *The Limits to Growth* (New York: Universe Books, 1972); Earl Cook, "Limits to Exploitation of Non-Renewable Resources," *Science* (February 20, 1976).
3. *The Global 2000 Report to the President* (Washington, D.C.: U.S. Government Printing Office, 1980), p. 1.
4. Julian Simon and Herman Kahn, eds., *The Resourceful Earth* (Oxford: Basil Blackwell, 1984), p. 45.
5. *The Global 2000 Report to the President*, p. 1.
6. Willis Harman, *An Incomplete Guide to the Future* (New York: Norton, 1979), chap. 2.
7. See Barry Hughes, *World Futures: A Critical Analysis of Alternatives* (Baltimore: Johns Hopkins University Press, 1985).
8. See Peter Berger and Thomas Luckmann, *The Social Construction of Reality* (New York: Doubleday, 1966).
9. See Konrad Waddington, *The Ethical Animal* (Chicago: University of Chicago Press, 1967), chap. 10; Donald Campbell, "Comments on the Sociobiology of Ethics and Moralizing," *Behavioral Science* (January 1979); Cyril Darlington, *The Evolution of Man and Society* (New York: Simon & Schuster, 1969).
10. For some notable exceptions see Nazli Choucri and Robert North, *Nations in Conflict* (San Francisco: W. H. Freeman, 1975); William Ophuls, *Ecology and the Politics of Scarcity* (San Francisco: W. H. Freeman, 1977); Harold Sprout and Margaret Sprout, *Toward a Politics of the Planet Earth* (New York: Van Nostrand Reinhold, 1971); Dennis Pirages, "The Ecological Perspective and the Social Sciences," *International Studies Quarterly* (September 1983).
11. An exception to this tendency is Gavin Boyd and Gerald Hopple, *Political Change and Foreign Policies* (London: Frances Pinter, 1987).
12. Dennis Pirages, *Global Ecopolitics: A New Context for International Relations* (North Scituate, Mass.: Duxbury Press, 1978), chap. 1.
13. Peter Worsley, *The Three Worlds: Culture and World Development* (Chicago: University of Chicago Press, 1984), pp. 42–43.
14. See Harold and Margaret Sprout, *The Ecological Perspective in Human Affairs* (Princeton, N.J.: Princeton University Press, 1965).
15. See William Leiss, *The Domination of Nature* (Boston: Beacon Press, 1972).
16. Riley Dunlap, "Paradigmatic Change in the Social Sciences: The Decline of Human Exemptionalism and the Emergence of an Ecological Paradigm," *American Behavioral Scientist* (September/October 1980).

17. See Herman Daly, *Steady-State Economics* (San Francisco: W. H. Freeman, 1977); see also the special issue of *American Behavioral Scientist* (September/October 1980).
18. Marvin Harris, *Cultural Materialism: The Struggle for a Science of Culture* (New York: Random House, 1979), p. 57.
19. Ibid., p. 52.
20. Ibid., p. 57.
21. See Marion Blute, "Sociocultural Evolution: An Untried Theory," *Behavioral Science* (January 1979).
22. See Kenneth Boulding, *Ecodynamics: A New Theory of Societal Evolution* (Beverly Hills: Sage, 1978).
23. Milton Rokeach, *The Nature of Human Values* (New York: Free Press, 1973), pp. 7–8.
24. See Donald Campbell, "The Conflict Between Social and Biological Evolution and the Concept of Original Sin," *Zygon* (September 1975).
25. "1985 World Population Data Sheet" (Washington, D.C.: Population Reference Bureau, 1985).
26. For example, see *World Resources 1987* (New York: Basic Books, 1987); *The State of the Environment 1985* (Paris: OECD, 1985).
27. See Willis Harman, op. cit.; Alvin Toffler, *The Third Wave* (New York: Morrow, 1980); Daniel Bell, *The Coming of Post-Industrial Society* (New York: Basic Books, 1973); Peter Drucker, *The Age of Discontinuity* (New York: Harper & Row, 1968).
28. The role of social surplus in growth and development is discussed in A. S. Boughey, "Environmental Crises—Past and Present," in Lester Bilsky, ed., *Historical Ecology* (Port Washington, N.Y.: Kennikat Press, 1980).
29. Earl Cook, *Man, Energy, Society* (San Francisco: W. H. Freeman, 1976), p. 19.
30. The dominant social paradigm concept was developed by Willis Harman, op. cit., chap. 2; see also Dennis Pirages and Paul Ehrlich, *Ark II: Social Response to Environmental Imperatives* (New York: Viking, 1974), chap. 2.
31. Thomas Kuhn, *The Structure of Scientific Revolutions* (Chicago: University of Chicago Press, 1962), chap. 5.
32. Ibid., chap. 7.
33. See Dennis Pirages and Paul Ehrlich, op. cit., p. 43.
34. Milton Rokeach, "From Individual to Institutional Values: With Special Reference to the Values of Science," in Milton Rokeach, ed., *Understanding Human Values* (New York: Free Press, 1979), pp. 47–51.
35. Willis Harman, op. cit., pp. 25–28.
36. Willis Harman, "The Coming Transformation," *The Futurist* (February 1977).
37. The World Bank, *World Development Report 1983* (New York: Oxford University Press, 1983), p. 7.
38. Edward Renshaw, *The End of Progress* (North Scituate, Mass.: Duxbury Press, 1976).
39. Orio Giarini and Henri Loubergé, *The Diminishing Returns of Technology* (New York: Pergamon Press, 1978).
40. See, for example, Christopher Freeman et al., *Unemployment and Technological Innovation: A Study of Long Waves and Economic Development* (Westport, Conn.: Greenwood Press, 1982).
41. See Alvin Toffler, op. cit.
42. Immanuel Wallerstein, *The Modern World System* (New York: Academic Press, 1974), pp. 33–37.
43. Ibid., pp. 38–39.
44. Ibid., pp. 39–48.
45. See Reid Bryson and Thomas Murray, *Climates of Hunger* (Madison: University of Wisconsin Press, 1977), chap. 6.

46. See Peter Worsley, op. cit., pp. 1–16.
47. David Finlay and Thomas Hovet, Jr., *International Relations on the Planet Earth* (New York: Harper & Row, 1975), pp. 22–23.
48. Nazli Choucri and Robert North, op. cit., p. 16.
49. Ibid., p. 15.
50. See Immanuel Wallerstein, *The Capitalist World Economy* (Cambridge: Cambridge University Press, 1976).
51. Richard Mansbach and John Vasquez, *In Search of Theory: A New Paradigm for Global Politics* (New York: Columbia University Press, 1981), p. 5.
52. Ibid., p. 5.
53. See ibid., chap. 1.
54. Hans Morgenthau, *Politics Among Nations* (New York: Knopf, 1948), chap. 9.

▼ Suggested Reading

DANIEL BELL, *The Coming of Post-Industrial Society* (New York: Basic Books, 1973).

NAZLI CHOUCRI AND ROBERT NORTH, *Nations in Conflict* (San Francisco: W. H. Freeman, 1975).

ALFRED CROSBY, *Ecological Imperialism and the Biological Expansion of Europe, 900–1900* (Cambridge: Cambridge University Press, 1986).

GIOVANNI DOSI, *Technical Change and Industrial Transformation* (New York: St. Martin's Press, 1984).

CHRISTOPHER FREEMAN ET AL., *Unemployment and Technological Innovation: A Study of Long Waves and Economic Development* (Westport, Conn.: Greenwood Press, 1982).

THEODORE GEIGER, *The Future of the International System: The United States and the World Political Economy* (Winchester, Ma.: Allen and Unwin, 1988).

ORIO GIARINI AND HENRI LOUBERGÉ, *The Diminishing Returns of Technology* (New York: Pergamon Press, 1978).

PETER GOUREVITCH, *Politics in Hard Times* (Ithaca, N.Y.: Cornell University Press, 1987).

WILLIS HARMAN, *An Incomplete Guide to the Future* (New York: W. W. Norton, 1979).

MARVIN HARRIS, *Cultural Materialism: The Struggle for a Science of Culture* (New York: Random House, 1979).

K. J. HOLSTI, *The Dividing Discipline: Hegemony and Diversity in International Theory* (Winchester, Mass.: Allen and Unwin., 1987).

ROBERT KEOHANE, *After Hegemony: Cooperation and Discord in the World Political Economy* (Princeton, N.J.: Princeton University Press, 1984).

ANDREW MAGUIRE AND JANET BROWN, eds., *Bordering on Trouble: Resources and Politics in Latin America* (Bethesda, Md.: Adler and Adler, 1986).

RICHARD MANSBACH AND JOHN VASQUEZ, *In Search of Theory: A New Paradigm for Global Politics* (New York: Columbia University Press, 1981).

WILLIAM MCNEILL, *Plagues and Peoples* (Garden City, N.Y.: Anchor Press, 1976).

WILLIAM MCNEILL, *The Pursuit of Power: Technology, Armed Force and Society Since A.D. 1000* (Chicago: University of Chicago Press, 1982).

WILLIAM OPHULS, *Ecology and the Politics of Scarcity* (San Francisco: W. H. Freeman, 1977).

MICHAEL PIORE AND CHARLES SABEL, *The Second Industrial Divide* (New York: Basic Books, 1984).

ANDREW SCOTT, *The Dynamics of Interdependence* (Chapel Hill, N.C.: University of North Carolina Press, 1982).

ALVIN TOFFLER, *The Third Wave* (New York: William Morrow, 1980).

IMMANUEL WALLERSTEIN, *The Modern World System* (New York: Academic Press, 1974).

ANDERS WIJKMAN AND LLOYD TIMBERLAKE, *Natural Disasters: Acts of God or Acts of Man* (London: Earthscan, 1984).

2

The Changing Global Context

Considerable evidence shows that the context within which relations among nations take place is rapidly changing, driving a shift away from the politics of the expansionist, materialist industrial revolution toward some type of post-industrial, less resource-intensive alternative. The last two decades have been filled with anomalies and crises indicative of an international system under stress. Economically, this period has seen two energy crisis cycles with accompanying price fluctuations, a major world food shortage that many thought would lead to global starvation, two very deep global recessions, a protracted global debt problem, and a collapse of world stock markets. In the decade prior to 1973, the annual real growth rate of the industrial economies was about 5 percent. Since then it has been in the vicinity of 2 percent. The less developed countries saw their vigorous growth rate of 6 percent of the previous decade drop to less than 2 percent in the first half of the 1980s. Even the oil-exporting countries, which did so well in the 1970s, were shocked when the bottom abruptly fell out of the oil market in the mid 1980s.[1]

Politically, the international power hierarchy is in disarray as the United States has moved from its status as the world's largest creditor nation to being the largest debtor. No longer is any hegemonic power capable of organizing the international system; it is now characterized by an uneasy pluralism. Nations that were once nearly powerless in international affairs have emerged as significant actors. The Group of 77, a collection of many of the less developed countries that now numbers nearly 100, still presses demands for the establishment of a new international economic order, and several debtor countries, such as Peru, Mexico, and Brazil, have informed multinational banks that principal and interest due on their loans will be repaid at the rates and times these countries find appropriate. And terrorists of many stripes seem able to exploit the weaknesses of complex technological societies with regularity.

Even in the global socioculture sphere, resistance to the continued spread of industrial modernization seems to be growing. A resurgence of Islamic fundamentalism is taking place on a broad front. Initially manifest in the reversal of modernization in Iran, the sudden surge of Islam has spawned political instability in the Middle East and social conflict in many countries where Islam and other religions previously coexisted. Outside the Islamic sphere of influence, other traditional forces, including fundamentalist Christians in the United States, have begun to attack the basic principles of the secular industrial revolution. On a slightly different note, many less developed countries have banded together to demand a new world information order that would correct the perceived cultural distortions coming out of the Western media.

The theoretical framework and ideas developed in the first chapter can be employed in an anticipatory mode to attempt to make sense of what seems to be an increasingly topsy-turvy world associated with discontinuity and instability that is accompanying the transition from second-wave industrial to third-wave post-industrial international politics.[2] Strong ecological currents are pushing and technological factors are pulling the international system and its component parts into unknown territory. The push factors that are shaping the politics of the twenty-first century are part of an interconnected complex of population and resource problems that has been referred to as the *global problematique*.[3] These include discontinuities in population, ranging from the population bomb that is ticking in many of the less developed countries to the graying of many of the industrial countries that now are approaching zero population growth, as well as pressures on natural resources associated with growing populations and industrial development. Recent boom-and-bust cycles of basic commodity prices attest to underlying pressures on resources and lack of adequate spare production capacity when the global economy moves into periods of significant growth. Among the primary pull factors shaping a new domestic and international political agenda is a generation of non-resource-intensive technologies in telecommunications, information processing, and biology that is both opening up new opportunities for the human race and raising significant domestic and international policy issues.

▼ Demographic Discontinuities

Large-scale demographic changes are important ecological factors creating discontinuities and shaping new domestic and international political and economic issues. On a global scale the widely recognized population problem is one of rapid exponential growth with very short doubling times. In the year 1650 there were only about 500 million people living on the face of the earth. It took nearly two hundred years for world population to double, reaching a figure of about 1 billion by the year 1850. But the next doubling of the world's population took place in only eighty years, yielding two billion human beings by 1930. In 1970 world population stood at about 4 billion, this doubling having taken place in only forty years. The present population of the world is pressing close to 5 billion. Demographers estimate that it is growing at a little more than 1.7 percent each year. At this rate of growth the world's population would nearly double again in another forty years.[4] Because of major efforts made prior to the 1980s in the family planning field, particularly in China with its massive population of more than 1.1 billion, the rate of global population increase will probably slow somewhat, yielding a figure of about 8 billion shortly after the year 2020.

No one really knows how many people the earth can ultimately support, and it certainly would be very unpleasant to find out. Some claim that a global population of 20 to 30 billion could be accommodated while others argue that a population overshoot has already taken place.[5] Much depends on assumptions made about the standards of living that these hypothetical masses would experience. A continuation of existing global patterns of population growth means that most new additions would occur in the less developed world, which already is undergoing significant ecological stress. The pressure of larger numbers of people and their livestock on the land has accelerated erosion and desertification in many parts of Africa and Asia, and recent famines there are a sign that populations have moved well beyond the long-term carrying capacity of the land.[6]

Leaving aside fascinating but somewhat esoteric speculation about the ultimate number of human beings that could be supported by the global ecosystem, four more pressing kinds of demographic changes are creating problems in the impacted countries and can be expected to cause friction in future relations among them. Primary among these is a cluster of issues related to the population explosion occurring in many of the less developed countries. On the other side of the demographic coin, most of the mature industrial countries face a far different agenda of problems related to the graying of their societies, a result of an increase in average age due to a drop in birth rates. Large-scale migration of people both within and among countries is a third factor causing persistent problems. Finally, differential rates of population growth both within and among countries suggest yet another agenda of potential conflicts.

The origins of the population boom in many of the less developed countries are found in the uneven penetration of the modernization process accompanying the outward spread of the industrial revolution. Modernization has many different aspects, but in keeping with the framework developed in Chapter 1, they can be grouped within techno-ecological, structural, and value domains. Industrialization means the introduction of new technologies, and the process initially gains widespread acceptance in developing countries because it creates jobs and is associated with obvious material benefits. Following the successful penetration by industrial technologies, however, deeper modernization requires structural and value changes less readily accepted by the traditional elements in a developing society. Full-scale modernization requires new political, economic, and social institutions if traditional patterns of behavior are to be changed and feudal values are to be transformed. But in many cases people in these countries are "rooted in traditional village agriculture, locked in near-feudal landholding patterns, [and] dominated by self-serving elites desperate to preserve their power. . . ."[7] In these environments, institutional and psychological changes take place very slowly, sometimes only over generations. Thus, in many developing countries there is significant turmoil and violence as advocates of traditional ways of looking at things clash with proponents of a new industrial world view.

The linkage between industrial modernization and rapid population growth is explained by the concept of the demographic transition. It refers to demographic regularities associated with phases of the modernization process. These regularities, reflected in movement through three stages, result from the interplay between industrialization and the introduction of related life-prolonging technologies and the pronatalist institutions and values that dominate agrarian societies. The first and last of the stages of the demographic transition are characterized by relative population equilibrium. It is during the middle stage that rapid population growth takes place.[8]

Populations in pre-industrial societies are characterized by relatively high birth rates. But since infant mortality is also high and life expectancy is short, population equilibrium is maintained. This type of stability results from congruence between long-existing demographic realities, high infant mortality and death rates, and the traditional pronatalist norms that condition reproductive behavior. Large families are considered natural, and children are valued for many reasons, not least of which is that they provide parents a certain "social security" in old age.[9]

The rapid population growth being experienced in much of the world occurs in the second stage of the demographic transition and follows from the introduction of disease-controlling and life-extending technologies. The reproduction norms characteristic of agrarian societies persist in a changed world where the nutritional and medical components of modernization begin to have a significant impact on infant mortality and life expectancy. More children survive to become adults and they live longer, but pre-industrial birth rates persist. Table 2-1 indicates the situation in some of the countries experiencing the most acute population increases. In Kenya, for example, the crude birth rate is now fifty-four per thousand, but the death rate has dropped to thirteen per thousand. This yields a population growth rate of 4.1 percent per year. At this rate, Kenya's population will double in only seventeen years, a staggering prospect for a country already facing massive economic development problems. Much of the less developed world faces similar demographic problems, although in many cases they are not quite as stark as those of Kenya.

These population growth rates will result in mammoth economic, political, and social problems in many of these countries during the next century. But the problem is not only one of current birth rates; it's also one of demographic momentum pent up in young populations. Even if by some miracle people could be persuaded to cut back to a two-child family in these countries over the next decade, substantial population increase is inevitable because of the large numbers of young people in the populations. In the industrialized world, about 22 percent of the population is under the age of 15. In Kenya, by contrast, over half of the population falls into this age category.[10]

A major effort to defuse the population boom has been under way for more than two decades and has had mixed success. It appears that the world passed through the most rapid rate of population growth in the late 1970s, when it hovered just above

Table 2-1 Rapidly growing countries.

Country	Annual Increase (%)	Crude Birth Rate*	Crude Death Rate*	% Population Under 15
Kenya	4.1	54	13	51
Syria	3.8	47	9	49
Rwanda	3.7	53	16	48
Zambia	3.7	50	13	47
Jordan	3.6	42	6	51
Tanzania	3.6	50	15	48
Zimbabwe	3.5	47	12	48
Nicaragua	3.5	43	8	47
Iraq	3.5	45	10	48
Congo	3.4	47	13	45
North Yemen	3.4	53	19	49
Uganda	3.4	50	16	48
Botswana	3.4	48	14	48

*Per thousand; all countries with populations less than one million not considered; slight errors due to rounding.
SOURCE: "1988 World Population Data Sheet," Population Reference Bureau.

2 percent per year. Since then, the estimated rate of growth has dropped to about 1.7 percent, but this small retreat from peak growth has been accompanied by complacency and an important reversal of U.S. population policies. A significant part of the international family planning effort has been financed and supported by the U.S. government, which prior to the Reagan administration contributed much to resolving world population problems. In the summer of 1984, however, the U.S. government reversed its traditional position in support of family planning at the World Population Conference in Mexico City. U.S. delegates took the position that stimulating free markets, which they claimed would lead to almost automatic economic progress in the less developed countries, would cause a natural decline in population growth rates.[11] This return to laissez faire policies by the Reagan administration, in response to fundamentalist religious pressures, represented backsliding that could well undo decades of hard work, which have led to the slight decrease in world population growth rates.

The Chinese role in changing the world demographic outlook is also an important factor to consider when assessing the recent decline from peak rates of population growth. The 1.1 billion Chinese in the People's Republic of China make up nearly one quarter of the world's population, and the seeming success of the Chinese one-child family policy is responsible for a large share of the worldwide decline. The rate of population growth in China is now slightly above 1 percent per year, a staggering achievement given the rapid rates of growth in China in the 1960s.[12] But this change has been accomplished only with the continued pressure of a major government campaign—including the use of economic rewards and sanctions—aimed at changing attitudes toward large families. The major role of the government in turning the Chinese situation around has been disturbing to many liberals, who may be worried about world population growth and like the results in China but don't necessarily approve of some of the methods the Chinese used to convince people to stop after having the first child. Pressures exist both within and outside of China for a softening of existing population policies, and there are indications that the population growth rate is increasing again as a result.[13]

Many exclusionists take the position that future economic growth in the areas of the world now experiencing the most rapid rates of population growth will push them naturally into the third stage of the demographic transition, cut birth rates, and restore population equilibrium. Given the dismal economic records of many of these countries over the last two decades, however, such a transition, without strong and effective family planning policies leading the way, would seem to be wishful thinking. It is no accident, for example, that the economically stagnant revolutionary cauldron in Central America is supported by some of the world's highest population growth rates. Populations in the region as a whole are growing at 2.6 percent per year, but in several of these countries the rate of growth is in excess of 3 percent per year.[14] Given dismal economic prospects for this and other regions, a large army of the unemployed and underemployed can be expected to foment continued political unrest.[15]

The uneven spread of the industrial revolution has created differential economic opportunities and has thus sparked large-scale movements of people within and among nations. Within most industrializing countries a surge of population is moving from the countryside to cities, as job opportunities resulting from industrial growth act as a magnet pulling people to urban areas. Differences of economic opportunity among countries are a powerful stimulus for both legal and illegal international migration. And political conflicts in various parts of the world have created and will continue to create a growing population of refugees, people without homelands and limited hope for resettlement.

Rural–urban migration, rampant in most industrializing countries, is expected to continue well into the twenty-first century, creating megametropolises that will be increasingly difficult to supply, service, and govern. In 1950, less than 8 percent of the world's population lived in cities of more than one million people. By the mid 1980s, this figure had more than doubled to 17 percent. It is projected that 27 percent of the world's population will live in such cities by the year 2020.[16] If current migration trends continue, the nineteen million people currently in Mexico City will explode to thirty-nine million by the year 2034, a crush of humanity that will overpower the already stressed sanitation and transportation infrastructures. At the same time, eight of the world's cities will have populations of more than twenty-five million, larger than any existing cities, and they all will be located in economically less developed countries.[17] This ecological chaos will certainly be reflected in urban social and political instability as a restive mob of urban unemployed continues to grow. The threat of unrest in many contemporary cities in less developed countries already has led governments to put economically questionable commodity subsidies in place to keep prices low and city dwellers pacified. It will be increasingly difficult, if not impossible, to fund larger quantities of subsidized commodities as cities grow in size, possibly leading to new urban violence when steps are taken to curb the heavy subsidies.

A related negative aspect of rapid urbanization is the resulting neglect of the agricultural sector. In one developing country after another, government policies favoring rapid industrialization and urbanization have had a devastating impact on agricultural productivity, turning countries that were once nearly self-sufficient in agriculture into major importers of agricultural commodities. Nigeria, for example, was once self-sufficient in food, but industrialization and the flood of oil money into Nigeria in the 1970s led the country to neglect agriculture and created significant food deficits in the 1980s.[18] These urbanization problems are not inevitable. China, for example, has made significant headway in slowing the flow to urban areas by limiting migration and locating light industry in the countryside. But stemming the tide of urbanization in many other countries seems beyond the political vision or capabilities of current leaders.

An imbalance in levels of economic development and opportunity is fostering a similar burgeoning flow of people among countries. Demographic push factors are squeezing people out of densely populated countries in the second stage of the demographic transition, and economic pull factors are drawing them to more affluent countries. These push factors in migration arise from growing lateral pressure in less developed countries: growing populations in the face of limited opportunities. Many Third World leaders see migration as a safety valve that reduces internal political pressures. In the western hemisphere, growing populations in Mexico, Central America, and the Caribbean, pushed by lack of meaningful job opportunities, continue to flow into the United States, legally and illegally. This increasing flood of migrants has raised big political issues in the United States and is causing friction in relations between the United States and its neighbors to the south.[19] There is every reason to expect that the population explosion will continue to worsen conditions in these countries of limited economic opportunities and that the problems associated with the growing flow of legal and illegal migrants to wealthier countries will increase apace.

Movement of guest workers from poor to rich countries has also served to ease population pressures in some countries. Western Europe and the oil-rich countries of the Middle East have at various times played host to large numbers of unskilled and semiskilled workers. These workers have been permitted to labor for short periods and have remitted most of their earnings to families back home. But the global economic slowdown of the 1980s led to large-scale expulsions of guest workers and major problems of reabsorption in their home countries.[20]

Finally, there is now a floating population of more than 8 million that has fled revolution, turmoil, and political persecution. More than 5.8 million of these refugees are housed in camps in various parts of the Near East and Asia, 2 million refugees are found in Africa, and significant numbers subsist in East Asia and Latin America.[21] Continued political instability in many parts of the world has kept the number of refugees growing steadily over the last decade, and there is little reason to expect a reversal of this troubling trend in the near future.

While the outward spread of industrial modernization has led to destabilizing population growth and related migration problems in the less developed parts of the world, the nascent post-industrial revolution is giving rise to a series of quite different issues in the mature industrial countries. Many of these countries have moved well into the third stage of the demographic transition and have reached or shortly will reach zero population growth (ZPG). Denmark, West Germany, Austria, and Hungary have moved beyond ZPG and are experiencing declines in population.[22] This post-industrial type of demographic change is creating its own set of discontinuities, problems, and policy issues, which will be increasingly obvious in structural change and conflicts over the next century.

Figure 2-1 illustrates age distributions in typical industrializing (Mexico), industrialized (United States), and post-industrial (Germany) countries. The Mexican age pyramid illustrates problems faced by Third World countries attempting to cope with the fallout from the population bomb. Nearly 45 percent of the population is under the age of fifteen, creating demographic momentum and many of the development ills mentioned above. The age distributions for the United States and West Germany, however, more closely resemble rectangles, as dropping birth rates and extended

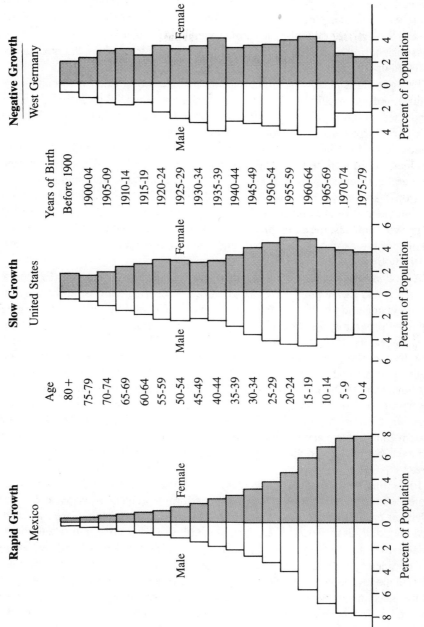

Figure 2-1 Population composition, age, and sex.

SOURCE: Population Reference Bureau

life expectancies combine to even out population distributions. This evening-out process means that, given current trends, the median age will continue to increase and the numbers in most age categories will become approximately equal. In West Germany and Sweden, for example, the median age of the population is now thirty-eight, contrasting sharply with many of the less developed countries where the median age is only about fifteen.[23]

This graying of mature industrial societies will likely be a factor in forcing future structural and value shifts. The dominant social paradigm of industrial societies depends heavily on rapid growth to sustain liberal structures and values. Without a return to vigorous pronatalist policies—which remains a future possibility—the expansionist way of life in the graying countries, and ultimately many industrial beliefs and values, could well be threatened. At minimum, major battles can be expected over liberal entitlement programs such as social security, as governments of aging, slow-growth societies begin to discover that there are real limits to the services that can be provided.[24] But beyond the direct and observable economic impact of graying slow-growth societies, the aging process could also be expected to lead to significant hardening of the psychological, social, and political arteries that might be manifest in a steady drift toward sociopolitical conservatism.

A look into the future of entitlement programs in the Organization for Economic Cooperation and Development (OECD) countries is sobering. Unless retirement concepts are changed dramatically, these economies could well be drained by existing entitlement programs, the products of the liberal value system that developed at advanced stages of industrial growth. This situation could also create an intolerable burden for the next generation of workers stuck with the crushing obligation of supporting an army of retired citizens. Because of the graying phenomenon in West Germany, for example, by the year 2030 there could be nearly one pension recipient for every worker contributing to the system. Although other industrial countries are in somewhat better shape than West Germany, in the United States and the United Kingdom each beneficiary could be supported by two and one-half workers and in Japan each beneficiary could be supported by only one and two-thirds workers in the year 2030.[25]

Projecting forward to the next century, unless there are major changes in policy designed to keep the aging economically active, reform retirement concepts, and curtail entitlement programs, the dynamism created by the social surplus of the industrial period may well disappear. This surplus, much of which has gone into social investment activities during the period of industrial expansion, could easily be swallowed up by the increasing maintenance costs of graying, liberal societies. For example, in 1960 in the seven major OECD countries, 49 percent of public social expenditures went for pensions and health care—essentially maintenance payments—whereas 30 percent of such expenditures went for education, basically an investment in future generations. By 1981, pensions and health care expenditures had reached 61 percent of public social expenditures and have steadily climbed since then. Educational expenditures, by contrast, had dropped to 20 percent and have continued to decline.[26]

A final demographic challenge to future global stability is created by differential population growth rates both within and among countries. Within countries composed of two or more major population groupings, growth rates of the faster growing segment will concern the segment growing more slowly and may well be a source of conflict. By the same token, the dramatically different demographic futures of industrializing and industrially mature countries may very well portend growing friction between them in a much more contentious international system. At present, approximately one quarter of the world's people live in the more developed part of the planet and the other three quarters live in the less developed countries. Ten percent of the world's population—about 500 million people—are North Americans and Western Europeans. Because of large differences in growth rates, these figures will rapidly change in the twenty-first century (see Table 2-2).

By the year 2020, the Western European and North American populations will have grown only slightly. But the 557 million people then living in these countries will have shrunk to only about 7 percent of the world's total, since the LDCs are expected to experience much more rapid population growth. By the year 2020, only 23 percent of the population in the more developed world will be under the age of fifteen, and 12 percent will be over the age of sixty-four. In the less developed world, by contrast, 39 percent of the population will be under the age of fifteen and only 4 percent will have reached retirement age. Although it is much more speculative to project out to the year 2100, demographers note that existing trends could shrink the combined American and Western European population still further, to about one twentieth of the world total.[27]

The impact of this growing demographic gap on future relations between a post-industrial North and an industrializing South remains a matter of conjecture, but if differences in age structures and rates of population growth are to be reflected in the future development of institutions and ideologies, serious tensions could be expected to grow out of this potentially volatile mix. Appreciation of and sympathy for the plight of the less developed countries waned in the mature industrial countries during the 1980s, and perhaps this is a harbinger of things to come in a much more polarized future world. In the post-industrial countries, which will make up

Table 2-2 World population projections, in millions.

	1987	2020	2100
Asia	2930	4584	5749
Africa	601	1479	2477
Oceania	25	35	40
Latin America	421	712	915
North America	270	326	341
Europe	495	502	547
USSR	284	355	376
World	5026	7992	10445

Source: "1986, 1987 World Population Data Sheets," Population Reference Bureau.

a much smaller portion of the world's future population, it is possible that graying populations will tenaciously cling to conservative values and unrealistic images of global reality. In the less developed world, by contrast, rapidly growing and restive young populations may feel much less bound by accepted conventions and traditions. Radical ideologies, new political movements, or even terrorism could grow out of the frustrations that may face a generation of young people in the South who see little reason to live by rules set by reactionaries in the North.

▼ Natural Resources: Growing Limitations

The rapid increase in world population and significant resource-intensive industrial growth still taking place in much of the world have combined to place considerable stress on the supply of natural resources essential for human well-being. In a formal sense, "a resource is anything needed by an organism, population, or ecosystem which, by its increasing availability up to an optimal or sufficient level, allows an increasing rate of energy conversion."[28] The global ecosystem supplies a variety of conventional resources such as air, water, energy, food, and minerals. Some resources, such as energy from the sun, are renewable and cannot be depleted by human activities. Others, such as fossil fuels, are nonrenewable and can only be used once. But there are also important nonconventional resources, such as the waste dispersal capabilities of the atmosphere and hydrosphere, that are often neglected in assessments of resource adequacy. Since the onset of the first oil shock, concern has grown that the global supply of various kinds of natural resources is inadequate to meet projected future human needs and thus presents a major challenge to future economic growth.

Human populations have always competed with other species for resources and have periodically expanded to run up against the resource limitations of the territories they have occupied. For example, in agrarian societies during times of relatively good weather for growing crops human populations have tended to expand rapidly, being cut back by nature during less favorable periods.[29] These natural fluctuations that have kept the size of human populations in check have been perceived by those affected as tragedies, famines, and plagues.

During the worldwide spread of the industrial revolution, obtaining fossil fuels and nonfuel minerals became major security considerations for growing nations. Some of these countries went through the industrialization process with very few indigenous resources, acquiring what was needed from other countries. Japan, for example, has been very dependent on foreign sources of natural resources, maintaining access to them first by developing political and military spheres of influence in Asia and more recently by being actively involved in international trade and resource diplomacy. Much of Western Europe passed through early industrialization using ample coal reserves, but, aside from North Sea reserves, there is now relatively little petroleum and natural gas in the region. The United States was endowed with copious quantities of natural resources, but decades of industrial growth have led to growing dependence on other nations for many critical resources, including a significant portion of the oil consumed. Of all of the major industrialized powers, only the Soviet Union can claim across-the-board resource self-sufficiency.

Nations employ various strategies to deal with resource constraints, ranging from technological innovations, which both create and conserve natural resources, to international trade, which can mutually enhance resources available to trading partners. Acquiring resources from abroad, however, is not without risk. In an increasingly interdependent world, closer trade relations among nations have enhanced sensitivity to developments in partner countries. But as resource constraints become more serious, as in the case of Japan, unwanted vulnerabilities can develop that inhibit foreign policy autonomy.[30] During the first energy crisis, for example, Japan was nearly 100 percent dependent on foreign suppliers for the petroleum that was so vital to the Japanese economy. Japan's actions then and subsequent Japanese policies in the Middle East have been very much constrained by the necessity of maintaining good relations with petroleum suppliers.

The extent to which future shortages of fuels and minerals will be factors creating problems and discontinuities in international relations is controversial. On the one hand, the inclusionist perspective sees rapid population growth and industrial development in various parts of the world to be overwhelming the capability of the global ecosystem to provide resources and environmental services essential to future human progress. On the other hand, exclusionists see little reason for concern and have responded with a blitz of books and articles touting the virtues of human initiative expressed in new technologies that will create a future world where abundant resources will sustain a much larger world population.

Concern about adequacy of the world's conventional and nonconventional resource base in the face of rapidly rising material demands was expressed well before the onset of the first oil crisis. Two years earlier a group of researchers working under the sponsorship of the Club of Rome—a private international organization made up of influential individuals concerned about the planetary future—published the results of a computer simulation of global futures that revealed a developing "global problematique" of interrelated population and resource issues that defied simple solutions. The first energy crisis, coming on the heels of the publication of the report, was seen by many as a validation of the limits-to-growth thesis and thus a harbinger of resource conflicts to come. The scarcity theme was picked up politically by the Carter administration, which commissioned a massive interagency study of future global resource problems.

The group that coordinated the *Global 2000* study concluded its evaluation of population and resource trends with a letter to the president, warning,

> Environmental, resource and population stresses are intensifying and will increasingly determine the quality of human life on our planet. These stresses are already severe enough to deny many millions of people basic needs for food, shelter, health, and jobs, or any hope of betterment. At the same time, the Earth's carrying capacity—the ability of biological systems to provide resources for human needs—is eroding. The trends reflected in the *Global 2000* suggest strongly a progressive degradation and impoverishment of the earth's natural resource base.[31]

The exclusionist technocrat opposition was quick to counterattack. Supported by a new administration that was basically opposed to long-term assessments, planning, and interfering with market mechanisms, the exclusionists argued that, for the bulk of the human race, things had never been better. Exclusionist Julian Simon played down the seriousness of world population problems, arguing that people are the world's ultimate resource. He further argued history shows that during the industrial revolution the real prices of basic commodities actually came down, indicating little evidence of impending scarcities.[32] Simon and premiere technological optimist Herman Kahn, with support from the conservative Heritage Foundation, commissioned an anti–*Global 2000* report, *The Resourceful Earth*. This book, dedicated to refuting the conclusions of the *Global 2000* study, concluded that

> environmental, resource and population stresses are diminishing, and with the passage of time will have less influence than now upon the quality of human life on our planet. These stresses have in the past always caused many people to suffer from lack of food, shelter, health, and jobs, but the trend is toward less rather than more of such suffering. Especially important and noteworthy is the dramatic trend toward longer and healthier life throughout all the world. Because of increases in knowledge, the Earth's "carrying capacity" has been increasing throughout the decades and centuries and millenia to such an extent that the term "carrying capacity" has by now no useful meaning.[33]

Given these diametrically opposed world views, the exclusionists anchored firmly in extrapolation from the past and the inclusionists anticipating a much different future, what can be gleaned from these studies about the role of resources in future relations among nations? The exclusionists may be right to point out that technology has been an important factor in driving down the relative prices of many raw materials. It is also true that market mechanisms compensate for resource scarcities by driving up prices, thus leading to greater supply and diminished demand. But events of the last two decades—two oil crises, a food crisis, an explosion in the price of basic commodities, new discoveries about the extent of global air pollution, a global recession, a world debt crisis, and a stock market crash—all tend to support a somewhat less bullish position. As is usually the case in confrontations of this nature, the matter is complex and both sides have something significant to contribute.

Natural resources differ in their abundance and importance. Without sunlight and air all human beings would die. But if all the copper ore in the world were to disappear tomorrow, life would certainly go on. Energy, air, and water clearly are the most basic of the natural resources needed by human populations. All of them are limited in supply, but are essential for food production. Energy is important because it is a resource that can be used to transform other resources. Shortages of fresh water, for example, can be overcome by energy-intensive desalinization. World food production has been increased tremendously in recent years, but not by bringing more land under cultivation. Instead, each acre of farmed land has been made more productive through application of energy-intensive fertilizers. Water is important because of the role it plays in growing crops, but also because every human body requires at least a minimum amount of water on a regular basis to stay alive. And air is not only critical for human survival, but the integrity of the entire atmosphere must be maintained if potentially disastrous alterations of climate are to be avoided.

The nonfuel resources, on which much of the scarcity debate has focused, are of a different nature. Iron ore, bauxite, and a host of other minerals are fairly abundant in the earth's crust, and a remarkable amount of future industrialization would be required on a planetary scale in order to make depletion of them a major problem. If ample cheap energy should become available, new extraction technologies could undoubtedly hold down prices of most of these minerals for the foreseeable future. And if prices rise significantly, seabed mining is a potential source for new reserves of many of them. The catch, however, is that reserves of critical minerals are not randomly distributed among the countries of the world, and serious national security problems face resource-deficient countries seeking to maintain access to them in a future world facing tighter constraints.[34]

Both sides in this debate over natural resource adequacy miss the economically and politically destabilizing consequences of rapid changes in the human relationship to the global ecosystem. It is not absolute per capita quantities of resources potentially available to human beings that is most important. It is instead the capacity of natural resource systems to absorb rapidly fluctuating demands that is most crucial in avoiding dislocations. The two energy crises were not triggered by Malthusian depletion of the fossil fuel resource base, although perceptions of scarcity conditioned development of them. The crises developed because rapid energy-intensive growth taxed the capabilities of the existing petroleum production and distribution systems. In the tight markets that spawned these crises, political events such as the abdication of the Shah of Iran triggered panic throughout world markets. In the subsequent cooling of global economic growth, demand for resources declined and commodity prices collapsed.

But more serious depletion problems are looming on the distant horizon. Energy from fossil fuels is a key resource upon which industrial civilization depends. Natural gas and petroleum are preferred fuels because they can be burned cleanly and are essential to existing transportation systems. But there is considerable evidence, much of which is discussed in the next chapter, that world supplies are inadequate to support traditional types of industrial growth for another century. Furthermore, new reserves are most often found in inhospitable locations, requiring massive amounts of capital for drilling and transportation. The alternatives to petroleum and natural gas all have drawbacks associated with their use. Where environmental considerations are unimportant, coal is an abundant source of future energy. But there are very few places where environmental considerations aren't important. Nuclear power is hardly an energy panacea, given the Chernobyl disaster and the tremendous multibillion dollar costs of building reactors. Solar energy is an abundant energy source, but threatens to remain too expensive to be practical unless significant research and development are funded in this area.

In summary, assessing the adequacy of the natural resource base available to support future industrial growth is a complicated matter, and the extent to which scarcity of resources will push nations and the international system into some post-industrial future is unclear. Some resources, such as nonfuel minerals, are certainly economically abundant on a global scale for the foreseeable future. But certain countries may have economic and political problems maintaining access to them. Other resources, particularly fuels, are matters of deeper concern because of the loss of the natural stored subsidy for growth that is inexorably disappearing as the energy and economic costs of obtaining them significantly increase. But in the short term, the scarcity problem is really one of perceptions and politics.[35] Perceptions of future scarcity exaggerate price fluctuations that plague basic commodity markets. And the unequal worldwide distribution of critical resources makes them potential weapons in confrontations between those who have them and want to extract higher prices and those who need to import them and want to keep basic commodity prices low.

▼ Technology and the Shrinking Planet

Population discontinuities and resource constraints may be seen as sticks that are pushing a transformation of international relations, but technological innovations are the carrots that are pulling in new directions. One clear impact of the array of "technotronic" technologies now on the horizon will be to shrink further an already small planet.[36] Over the last twenty years, major innovations in transportation have fostered the movement of people and goods around the world at unprecedented speed and decreasing cost. Developments in telecommunications, including the use of satellites and fiber-optic cables, will further accelerate this trend toward planetary togetherness and interdependence. Communications satellites have already made nearly instantaneous communication with once remote areas of the world a reality, and new fiber-optic cables being laid across the Atlantic Ocean will increase communications links between the United States and Europe many times over. This accelerated global movement of people and information will create new types of global interdependence ranging from worldwide financial markets to the rapid international spread of various diseases.

Post-industrial technologies are developing on many fronts, bringing with them exciting new prospects but also potential future problems. New communication capabilities, for example, already gave rise to the first internationally broadcast airplane hijacking in the Middle East. The Marcos–Aquino confrontation in the Philippines marked the first time that participants in a revolution gave internationally televised interviews as mobs were moving to the palace gates. On the other hand, global teleconferencing among world leaders and global news coverage enhance possibilities for world peace. Foreign leaders are now regularly given access to U.S. television audiences through live speeches and interviews broadcast via satellite. But it remains to be seen whether the telecommunications revolution will actually create greater international understanding or whether global familiarity will eventually breed mutual contempt.

Developments in other frontier areas of technology will create similar opportunities and problems. A revolution in computers and information processing opens up the prospect of increasing social knowledge, but it also raises the Orwellian specter of unwarranted government intrusion into personal matters. Internationally, the information revolution is giving rise to demands for a new international information order by less developed countries seeking protection from the impact of an unrestricted worldwide information flow. New biomedical technologies promise to extend the human life span and give human beings unprecedented control over reproductive and evolutionary processes. But these same technologies carry with them the risk of unforeseen environmental consequences of genetic engineering experiments as well as a host of deep moral issues regarding the sanctity of human life. Finally, the push into outer space carries with it the possibility of enhanced international cooperation in this realm but also the threat of a runaway "Star Wars" arms race, the cost of which could be beyond any weapons programs yet imagined.

It is useful to think of the issues arising from these relationships between technological change and international politics as being important in three different time frames. A set of long-term issues arises from the fact that the international power hierarchy has traditionally been maintained by national differences in technological sophistication. Long-term issues arise not only out of the obvious link between technology and the spread of sophisticated weaponry but, increasingly, out of less familiar concerns such as growing economic, political, and cultural interdependence among nations, "commons" problems of both a physical and technological nature, questions of controlling the spread and impact of certain cutting-edge technologies, and the significant and persistent gap between technologically sophisticated and less sophisticated countries. There is also an agenda of policy issues important in the medium term, clustering around high-technology competition. Competition questions include optimal patterns of investment in research and development, the role of the state in industrial policy, and the determination of rules for conduct in a period of intensifying international trade competition. Finally, a set of more immediate, contemporary questions focuses on the appropriateness of certain kinds of technology transfers in the East–West, North–South, and North–North directions.[37]

▼ Persistent Global Issues

Technology has always been a critical factor in creating and sustaining a hierarchy of nations. Indeed, it is easy to make the case that military technologies have been critical in shaping the political map of the world since the time of the Roman Empire.[38] During the advanced stages of the industrial revolution, however, nonmilitary technologies have assumed a much more important role in shaping this international order. A large population and access to resources are the minimum requirements for major power status, but no nation can now aspire to that status unless it also has high levels of achievement in both military and civilian research and development.

The close link between technology, military power, and position in the hierarchy of nations has been obvious throughout history. Less well remarked, though of increasing importance, is the link between technology, economic power, and this hierarchy. Humankind's fascination with sophisticated weaponry has hardly diminished in recent years, but development of nuclear and thermonuclear weaponry and related delivery vehicles has made the major powers much more hesitant to use force in situations that could escalate into nuclear confrontations. Thus, smaller nations are now freer to engage in their own adventures without great fear of major power involvement.[39] More important, however, the extended period of relative calm in military relations among the great powers since the end of the Korean War has spawned intense technology-based international economic competition that is increasingly supplanting military force as a foreign policy priority.

The United States emerged from the second world war with unchallenged supremacy in military technology. In fact, the conclusion of the war was hastened by the first use of nuclear weapons by U.S. forces in Asia. Since that time, U.S. military policy has relied heavily on a technological edge to enhance the capabilities of or to replace human beings in military encounters. During the Korean War, superior U.S. technology permitted smaller U.S. forces to hold large masses of Chinese troops to a stalemate. In the Vietnam War precision-guided bombs launched from high-altitude bombers were used, with mixed success, to compensate for manpower deficiencies on the ground. In future military endeavors the United States is planning to rely on electronic battlefields, in an effort to reduce the human element in combat to a minimum.

This U.S. emphasis on a technological edge has also been important in postwar international economic competition. The United States emerged from the second world war clearly dominant over the international economic system. Japan and Germany had been totally drained by the war and were in no position to compete economically with the United States. The wartime allies were also in a weakened position because of their losses. The Soviet Union had been deeply damaged by the war and had destroyed much of its productive potential through a scorched-earth policy that denied assets to the advancing German armies. The French and the British also were left in dire economic straits. Thus, the United States alone among the postwar major powers emerged with its productive capability largely unscathed by the effects of the war. Given this situation, the United States went virtually unchallenged in shaping the rules that governed postwar international economic competition. Key U.S. industries—steel, automobiles, aircraft, farm equipment, and even nuclear energy—dominated world markets technologically, and the United States maintained control over the world military and economic hierarchies for the quarter of a century following the war.

Beginning in the early 1970s, however, the military, economic, and political hegemony of the United States came under attack from both old and new quarters, and former enemies and allies alike emerged from the aftermath of the war as strong competitors. While the war destroyed the industrial infrastructure of Japan and Germany, it also destroyed vested economic, political, and social interests and gave impetus to reconstruction efforts that eventually catapulted these two countries into serious technologically sophisticated economic competition with the United States.[40] The Soviet Union, emphasizing heavy industry and a military buildup, emerged as a major military competitor. And a number of newly industrializing countries, such as Korea, Taiwan, and Singapore, began to acquire new technologies and make their own impact in world markets that had previously been the domain of the United States. Thus, a worldwide spread of technological know-how has substantially lessened U.S. preeminence in the international military and economic hierarchies and has complicated the matter of maintaining predictability, order, and established rules of conduct in relations among nations.

A second long-term impact of technological innovation on international relations is a steady movement toward more ecological, structural, value, and policy interdependence. Ecological interdependence has been increased by the needs of most industrial countries to import large quantities of natural resources from the less developed countries and by the great impact of each major industrial country on the shared global ecosystem. Greater structural interdependence has followed closely behind as a worldwide division of labor has become a reality through development of world capital markets and cheap and rapid transportation. Even a new value interdependence is being facilitated by increased international contacts made possible by new communications technologies. Finally, events of the 1980s—particularly the global stock market crash—have repeatedly demonstrated increased policy interdependence, both among the mature industrialized nations themselves and between them and the less developed countries.

Increased technology-induced interdependence among nations has both positive and negative aspects. On the positive side there is significant potential for the development of greater international understanding, based on more frequent and (one hopes) more accurate communications. The growth of global markets promises to introduce a better division of labor and more efficiency into the global economy and eventually to give all countries a stake in future economic and political stability. But on the negative side, a global division of labor in a more integrated global economy is not necessarily equally beneficial to all parties. Growing interdependence can give rise to new vulnerabilities in political-economic relationships; witness the resource warfare that broke out in the world energy market in the 1970s. Furthermore, potential new economic dislocations in the future will much more readily be transmitted across national borders. Continued monitoring and management of complex interdependence will be required if the advantages of it are to outweigh the liabilities.

A related third set of long-term connections between technology and international issues revolves around collective goods or commons, problems that are becoming more serious as industrialization continues its worldwide spread. These issues arise out of various kinds of assaults on the commonly shared global ecosystem and result from what normally are considered to be the successes of the industrial revolution. But industrialization exacts a significant toll from the global environment because of toxic liquids, gases, and solids that are by-products of industrial processes. Although in common language the word *disposed* is used in talking about these by-products, most of them are simply *dispersed* into the atmosphere and hydrosphere shared by all nations. There are now only rudimentary international treaties and agreements proscribing pollution by the many nations that often treat the commons as a global septic tank. In the early stages of the spread of the industrial revolution, the impact of this pollution was not large enough to be noticeable on a global scale. But as more countries have industrialized, the deterioration of the global commons has become a pressing concern. In several areas commons problems have already become the subject of international research efforts. Numerous studies have projected a serious buildup of carbon dioxide gas in the atmosphere, resulting from combustion of fossil fuels, thus creating a sort of global greenhouse. Within this global greenhouse, continuation of current trends will result in a significant warming of the earth's atmosphere by the middle of the next century. Similarly, the ozone layer in the atmosphere, a resource that keeps much of the sun's penetrating ultraviolet radiation from striking the earth, is being weakened, attacked by continued use of various chlorofluorocarbons and similar chemicals. The continued breakdown of this protective layer would substantially increase skin cancer worldwide and have other serious effects on human health. New technological innovations also permit attacks on the few remaining geographic commons, such as the deep seabed and the antarctic, regions previously protected by barriers to development.[41]

The complex nature of some new technologies and the damage that could result from their improper use create a fourth set of international problems: managing the spread of potentially dangerous cutting-edge technologies. Some new technologies, such as innovations in telecommunications, pose fairly straightforward regulatory problems that can be resolved through diplomacy within existing international organizations. Restrictions on direct broadcast satellites, allocation of radio frequencies, control of commercial use of satellite transmissions, limitations on transborder information flow, and allocation of communications satellite locations in geosynchronous orbit fall into this negotiable category.

But other new technologies, such as those in nuclear science and biology, pose more difficult management problems. In their efforts to win larger shares of limited markets, some firms or countries may prove willing to sell inappropriate technologies or to take imprudent risks that could eventually lead to large-scale catastrophes. The transfer of such technologies into societies that do not have adequate knowledge or infrastructure to support and regulate them is a major international problem. In the nuclear field, for example, much of the knowledge and equipment required to operate a commercial reactor program can also be used to fabricate nuclear weapons. Competitive pressures or simple greed can induce sellers to transfer such sensitive technologies under the guise of peaceful nuclear power production to countries that may seek nuclear weapons. Furthermore, nuclear reactors that are improperly constructed or maintained in less developed countries could break down, with potential leakage of radioactive materials and resulting catastrophes for those living nearby. Before Ferdinand Marcos was removed from power in the Philippines, for example, construction was well under way on a nuclear reactor located dangerously near an active volcano. Aspects of biotechnology, particularly those associated with genetic engineering, are also creating technology transfer dilemmas. In addition to the risk of innocent mistakes during research or manufacturing that could release dangerous organisms into the environment, the potential for the spread of biological warfare capability is growing significantly as new biotechnologies are developed.[42]

A final, major international dilemma persists because of an uneven development and spread of new technologies. Although it is not easy to quantify technological innovation, it is generally agreed that more than 95 percent of all significant innovations take place in the industrialized world. This means that, over time, a technology gap has developed and endures between countries that industrialized early and less developed countries struggling to find a place in the contemporary international economy. The early industrializers have the capital and technological know-how to compete in a world of accelerating competition, whereas less developed countries generally have neither. Thus, a persistent technology-induced economic gap separates rich and poor countries in the international system, and the evidence indicates that the gap isn't diminishing.

This gap cannot be shrunk without dramatically altering contemporary trade and aid practices. Technological innovation leads to a wider array of cheaper goods and services available in the countries that are already industrialized, which in turn leads to domination of world markets by firms located in those countries. Since under present circumstances technology is rarely freely transferred from rich to poor countries, the less developed countries seem consigned to exporting agricultural commodities and minerals for the foreseeable future. In fact, on a per capita basis there has been negative growth in many of these countries, and they are sometimes described as being in the Fourth World, a group of countries that is no longer expected to develop by following the paths laid out by the early industrializers. Particularly in an era of rapidly rising expectations throughout the global village, the significant and often growing gap between rich and poor will sharpen conflicts between the technologically advanced and technologically limited countries.

▼ Pressing Policy Questions

In a more immediate time frame, technology is increasingly seen by leaders in industrial countries as a national resource that must be carefully husbanded in order to maintain economic and military competitiveness. Thus, more immediate science and technology policy issues are now frequently debated as matters of national security. Scientific discovery and technological innovation are not random processes: they can be nurtured by appropriate policies. Some countries invest more heavily than others in public and private research and development, and some countries seem to get better results from their investments. Although no country has developed a magic formula for ensuring technological competitiveness, some countries, such as Japan and Korea, do much better at competition than others do.

The remarkable trade successes of the Japanese combined with huge trade deficits in the United States have sparked an ongoing debate in the U.S. over the government's role in establishing industrial policies. In recent years the United States has regularly run trade deficits well over $150 billion annually and has moved from being a major creditor in the international economy to being the world's greatest net debtor nation. There is thus great concern about identifying the scientific, technological, and engineering factors that lead to success in international economic competition, so that the United States may preserve industries and jobs in the face of tougher foreign competition.[43]

More traditional national security concerns keep a number of trade and technology transfer questions on the more openly debated policy agenda. Primary among these is a series of East–West technology transfer issues. Given the American historical edge in military technology, there is a great deal of concern to keep knowledge and equipment with possible military applications from being sold to the Eastern bloc countries. Several expert commissions have attempted to deal with aspects of this problem over the years, but have not come up with definitive prescriptions.[44]

Definitions of and prescriptions for protecting sensitive information and hardware in the face of the huge trade deficit vary considerably. Hard-liners would deny the sale of any technologically sophisticated products or processes to the Eastern bloc on national security grounds regardless of economic consequences. Thus, in the mid 1980s U.S. customs officials interdicted Apple computers bound for Eastern Europe on the grounds that they represented a security threat. Little matter that Eastern bloc diplomats could purchase these computers at many retail outlets elsewhere in the Western world and easily ship them home. The hard-line position is normally supported by officials in the Department of Defense, who argue that the United States is locked in mortal combat with the Soviet adversary and that any transfer of technology will tend to benefit the Soviet military machine. State and Commerce department officials usually take a more moderate position, which recognizes that what beleaguered U.S. companies aren't permitted to sell to the Eastern bloc will probably be sold there by companies in other Western countries. These arguments not only persist within U.S. policy circles but also periodically flare up in relations among OECD countries, the Europeans generally being much more liberal in their approach to trade with the Soviet Union and Eastern Europe.

Quite apart from arguments over the advisability of embargoing certain technology transfers to the Soviet Union are some critical questions about the feasibility of such embargoes. Intense economic competition within the Western alliance leads to pressures for bending accepted agreements or even outright cheating. Attempts have been made to coordinate Western trade policies through an organization known as Cocom, but successes have been mixed. On a more basic level, the Reagan administration made a major effort to prevent the transfer even of scientific information to the Eastern bloc. This involved refusing to let non-Western scientists attend certain professional meetings as well as efforts to control activities of foreign students on U.S. campuses. None of these efforts was very successful, although East–West cooperation has been inhibited in many important areas of research.

Whereas friction over East–West trade and technology transfer is rooted in military security concerns, a similar set of West–West issues among the OECD countries themselves is more closely related to trade and economic security interests. Most OECD countries ostensibly subscribe to a free trade philosophy championed by the United States since World War II. In reality, however, many of these countries don't always play by the established rules nor interpret them in the same way. Thus, subtle and not-so-subtle nontariff barriers to trade are frequently used to gain an edge in the intensified global economic competition. The often inferior competitive position of U.S. industry and related loss of American jobs has led to repeated congressional cries for more protection against foreign products, even though such demands fly in the face of professed U.S. policy.

Finally, there is a set of North–South trade and technology transfer issues dealing with the appropriateness of certain technology transfers. Successful transfer and management of new technologies require certain capabilities on the part of the recipient countries. Nuclear power technologies, for example, are not appropriate for politically unstable less developed countries that cannot guarantee the long-term stability and vigilance required to successfully manage nuclear power programs. But competitive pressures have led to sales of nuclear plants to most countries able to afford them. Another set of transfer issues revolves around the apparent unwillingness of firms in the industrial world to develop and market technologies that are appropriate and useful in less industrialized settings. Because the less developed countries have little capital with which to influence research and development programs in the developed countries, few technological breakthroughs have immediate applicability to enhancing productivity in the less developed world.

▼ Perceptions of Global Issues

Contemporary international politics is characterized by various kinds of political continuities coexisting with growing pressures for change. The continuities are mainly provided by an established generation of political "realists" who have learned about how nations behave from history books and have a vested interest in defining problems in conventional terms. The world view that colors their perceptions is most frequently exclusionist, and their solutions to problems are generally defined in the terminology of established power politics. International problems are perceived to be caused by traditional maladies, such as the maliciousness of other countries and their power-seeking leaders driven by alien ideologies. Solutions to problems are most frequently found in the measured use of military force.

A post-industrial paradigm for international relations suggests a different approach to understanding and resolving international issues. It has been visible recently in the actions of some forward-looking leaders and gives more appropriate guidance in dealing with future global problems. Proponents of this post-industrial, macropolitical paradigm generally operate from a nascent inclusionist framework and understand the dynamic relationships among demographic, ecological, and technological changes and the expanding agenda of global issues.[45] But the clash between proponents of old and new world views, in both theory and practice, will undoubtedly continue for decades until the policy issues generated by the factors outlined in this chapter break through the generational lag that characterizes much current political thinking.

The twenty-first century will be dominated by global issues that require an enlightened inclusionist response. Long-term trends in population growth and mobility, resource availability, and technological innovation make increased global interdependence an unavoidable reality. The future will be one of increased international movement of people, information, ideas, capital, and natural resources. The international politics of the future will revolve around this new interdependence and the policy coordination required to manage it. The chapters that follow further develop this inclusionist framework by identifying and analyzing the most important of these emerging global issues and the many policy issues and management problems associated with them.

▼ Notes

1. Figures are derived from the World Bank, *World Development Report 1986* (New York: Oxford University Press, 1986), chap. 2.
2. See Alvin Toffler, *The Third Wave* (New York: Morrow, 1980).
3. The term *global problematique* is used to describe an interlocking set of environmental, population, and institutional problems first outlined in Donella Meadows et al., *The Limits to Growth* (New York: Universe Books, 1972).
4. Growth rate taken from "1987 World Population Data Sheet" (Washington, D.C.: Population Reference Bureau, 1987); historical figures are found in Paul Ehrlich et al., *Ecoscience: Population, Resources, Environment* (San Francisco: W. H. Freeman, 1977), p. 183.
5. S. Fred Singer, ed., *Is There an Optimum Level of Population?* (New York: McGraw-Hill, 1971); William Catton, *Overshoot: The Ecological Basis of Revolutionary Change* (Urbana: University of Illinois Press, 1980).
6. Thomas Goliber, "Sub-Saharan Africa: Population Pressures on Development," *Population Bulletin* (February 1985); Lester Brown and Edward Wolf, *Reversing Africa's Decline* (Washington, D.C.: Worldwatch Institute, June 1985).
7. Alex Inkeles and David Smith, *Becoming Modern: Individual Change in Six Developing Countries* (Cambridge, Mass.: Harvard University Press, 1974), p. 4.
8. See Robert McNamara, "Time Bomb or Myth: The Population Problem," *Foreign Affairs* (Summer 1984); Maurice Kirk, "The Return of Malthus?" *Futures* (April 1984); W. Penn Handwerker, "The Modern Demographic Transition: An Analysis of Subsistence Choices and Reproductive Consequences," *American Anthropologist* (June 1986); Nathan Keyfitz, *Population Change and Social Policy* (Cambridge, Mass.: Abt Books, 1982).
9. For example, when a mother of twenty-seven children in Brazil was asked what she thought of family planning, she said she would have those children all over again because only God can determine how many children a woman should have. Reported in "Brazil Tries Birth Control," *World Press Review* (February 1984).
10. Data taken from "1987 World Population Data Sheet."
11. Taken from the official "Policy Statement of the United States of America at the United Nations International Conference on Population (Second Session)," Mexico City (August 6–13, 1984).
12. Population figures for China are from "1987 World Population Data Sheet."
13. "China's Birth Rate Reported on Rise," *Population Today* (May 1987).
14. Figures taken from "1987 World Population Data Sheet." See also Thomas Merrick, "Population Pressures in Latin America," *Population Bulletin* (July 1986).
15. See Allen Otten, "Population Explosion Is a Threat to the Stability of Latin America," *The Wall Street Journal* (February 17, 1984).
16. *World Population: Fundamentals of Growth* (Washington, D.C.: Population Reference Bureau, 1984). See also George Beier, "Can Third World Cities Cope?" *Population Bulletin* (December 1976).

17. Leon Bouvier, "Planet Earth 1984–2034: A Demographic Vision," *Population Bulletin* (February 1984), p. 22.

18. For example, in agriculture trade in 1963 Nigeria had a surplus of $163 million; by 1979 the situation had turned to a trade deficit of $715 million (figures taken from United Nations Food and Agricultural Organization Yearbooks in appropriate years).

19. See Michael Teitelbaum, "Immigration, Refugees and Foreign Policy," *International Organization* (Summer 1984), and "Immigrants and Refugees," *Foreign Affairs* (Fall 1980).

20. These clashes are often violent in less developed countries as returning workers pose a threat to those who are employed at home.

21. U.S. Department of State, *Country Reports on the World Refugee Situation: Statistics* (Washington, D.C.: U.S. Government Printing Office, 1984), p. 2. See also Kathleen Newland, *Refugees: The New International Politics of Displacement* (Washington, D.C.: Worldwatch Institute, March 1981).

22. See Allan Carlson, "Depopulation Bomb: The Withering of the Western World," *The Washington Post* (April 13, 1986); Dirk Van de Kaa, "Europe's Second Demographic Transition," *Population Bulletin* (March 1987).

23. Leon Bouvier, op. cit., p. 15.

24. Murray Gendell, "Sweden Faces Zero Population Growth," *Population Bulletin* (June 1980); Ben Wattenberg and Karl Zinsmeister, "The Birth Dearth: The Geopolitical Consequences," *Public Opinion* (December/January 1986); Allan Kelley, "The Birth Dearth: Economic Consequences," *Public Opinion* (December/January 1986); Richard Rose and Guy Peters, *Can Governments Go Bankrupt?* (New York: Basic Books, 1978).

25. "The Future of Social Expenditure," *The OECD Observer* (January 1984). See also *Social Expenditure: 1960–1990—Its Growth and Control* (Paris: OECD, 1984).

26. "Pensions After 2000: A Granny Crisis is Coming," *The Economist* (May 19, 1984).

27. Estimates are from "1986 World Population Data Sheet" (Washington, D.C.: Population Reference Bureau, 1986).

28. Kenneth Watt, *Principles of Environmental Science* (New York: McGraw-Hill, 1973), p. 20.

29. See William McNeill, *Plagues and Peoples* (Garden City, N.Y.: Anchor Press, 1976); Anders Wijkman and Lloyd Timberlake, *Natural Disasters: Acts of God or Acts of Man* (London: Earthscan, 1984).

30. Robert Keohane and Joseph Nye, *Power and Interdependence* (Boston: Little, Brown, 1977), pp. 12–13.

31. *The Global 2000 Report to the President* (Washington, D.C.: U.S. Government Printing Office, 1980), p. 1.

32. Julian Simon, *The Ultimate Resource* (Princeton, N.J.: Princeton University Press, 1981), chap 1.

33. Julian Simon and Herman Kahn, *The Resourceful Earth* (Oxford: Basil Blackwell, 1984), p. 45.

34. See H. E. Goeller and A. Zucker, "Infinite Resources: The Ultimate Strategy," *Science* (February 3, 1984); Douglas Bohi and Michael Toman, "Understanding Renewable Resource Supply Behavior," *Science* (February 25, 1983).

35. Ted Gurr, "On the Political Consequences of Scarcity and Economic Decline," *International Studies Quarterly* (March 1985).

36. Zbigniew Brzezinski, *Between Two Ages: America's Role in the Technotronic Era* (New York: Viking, 1970).

37. See John Granger, *Technology and International Relations* (San Francisco: W. H. Freeman, 1979); Victor Basiuk, *Technology, World Politics and American Policy* (New York: Columbia University Press, 1977); Anne Keatley, ed., *Technological Frontiers and Foreign Relations* (Washington, D.C.: National Academy of Sciences, 1985).

38. See William McNeill, *The Pursuit of Power: Technology, Armed Force and Society Since A.D. 1000* (Chicago: University of Chicago Press, 1982).

39. Richard Rosecrance, *The Rise of the Trading State* (New York: Basic Books, 1986), chap. 9.
40. This argument parallels that of Mancur Olson, *The Rise and Decline of Nations* (New Haven: Yale University Press, 1982).
41. See Marvin Soroos, *Beyond Sovereignty: The Challenge of Global Policy* (Columbia, S.C.: University of South Carolina Press, 1986).
42. See Office of Technology Assessment, *Commercial Biotechnology: An International Analysis* (Washington, D.C.: U.S. Congress, 1984); Jonathan Tucker, "Gene Wars," *Foreign Policy* (Winter 1984/85).
43. See Lester Thurow, *The Zero Sum Solution* (New York: Simon & Schuster, 1985).
44. *Scientific Communication and National Security* (Washington, D.C.: National Academy of Sciences, 1982); Mitchel Wallerstein, "Scientific Communication and National Security in 1984," *Science* (May 4, 1984); National Academy of Sciences, *Balancing the National Interest: U.S. National Security, Export Controls and Global Economic Competition* (Washington, D.C.: National Academy Press, 1987).
45. Richard Sterling, *Macropolitics* (New York: Alfred Knopf, 1974), p. 6.

▼ Suggested Reading

LESTER BROWN, ET AL., *The State of the World 1988* (New York: W. W. Norton, 1988).

WILLIAM CATTON, *Overshoot: The Ecological Basis for Revolutionary Change* (Urbana, Ill.: University of Illinois Press, 1980).

DAVID DRAKAKIS-SMITH, *The Third World City* (New York: Methuen, 1987).

PAUL EHRLICH ET AL., *Ecoscience: Population, Resources, Environment* (San Francisco: W. H. Freeman, 1977).

PAUL EHRLICH AND ANNE EHRLICH, *Extinction: The Causes and Consequences of the Disappearance of Species* (New York: Random House, 1981).

ALLAN FINDLAY AND ANNE FINDLAY, *Population and Development in the Third World* (New York: Methuen, 1987).

JOHN GRANGER, *Technology and International Relations* (San Francisco: W. H. Freeman, 1979).

OTTO HIERONYMI, ed., *Technology and International Relations* (New York: St. Martin's Press, 1987).

BARRY HUGHES, *World Futures: A Critical Analysis of Alternatives* (Baltimore: Johns Hopkins University Press, 1985).

The Global 2000 Report to the President (Washington, D.C.: U.S. Government Printing Office, 1980).

INTERNATIONAL BANK FOR RECONSTRUCTION AND DEVELOPMENT, *Population Change and Economic Development* (New York: Oxford University Press, 1985).

NATHAN KEYFITZ, *Population Change and Social Policy* (Cambridge, Mass.: Abt Books, 1982).

DONELLA MEADOWS ET AL., *The Limits to Growth* (New York: Universe Books, 1972).

NORMAN MYERS, *The Sinking Ark: A New Look at the Problem of Disappearing Species* (New York: Pergamon Press, 1979).

A. F. K. ORGANSKI ET AL., *Births, Deaths, and Taxes* (Chicago: University of Chicago Press, 1984).

ROBERT REPETTO, ed., *The Global Possible: Resources, Development, and the New Century* (New Haven, Conn.: Yale University Press, 1985).

JULIAN SIMON, *The Ultimate Resource* (Princeton: Princeton University Press, 1981).

JULIAN SIMON AND HERMAN KAHN, *The Resourceful Earth* (Oxford: Basil Blackwell, 1984).

MARVIN SOROOS, *Beyond Sovereignty: The Challenge of Global Policy* (Columbia, S.C.: University of South Carolina Press, 1986).

MICHAEL TEITELBAUM AND JAY WINTER, *The Fear of Population Decline* (New York: Academic Press, 1985).

JOSEPH TULCHIN, *Habitat, Health, and Development: A New Way of Looking at Cities in the Third World* (Boulder, Colo.: Lynne Rienner Publisher, 1986).

ARTHUR WESTING, ed., *Global Resources and International Conflict* (Oxford: Oxford University Press, 1986).

World Commission on Environment and Development, *Our Common Future* (New York: Oxford University Press, 1987).

World Resources 1987 (New York: Basic Books, 1987).

3

The
Political
Ecology
of Energy

The continued spread of the resource-intensive industrial revolution has put pressure on world fossil fuel reserves that once seemed infinite in relation to demands for them. The two energy shock cycles of the last fifteen years have focused attention on the growing vulnerabilities of certain energy-deficient industrial countries, the long-term inadequacy of the world's fossil fuel energy base, and the possibility that supply and price instability could become a major barrier to industrial growth in the twenty-first century. Much of the machinery used in industry now runs on the fossil fuel energy that has been at the core of heavy industrial development. The fossil fuels energize factories, power mechanized farms, and support large-scale transportation networks. Historically the fossil fuels have been taken for granted, since they have been so cheap and easy to obtain. But the two energy shocks heightened awareness of the finite nature of these critical resources and introduced at least a temporary note of urgency into energy policy planning.

Over the long span of human history prior to the onset of the industrial revolution, current solar income—energy from the sun—was the source of energy for the evolution of human societies. Much of this solar energy was originally captured by photosynthetic processes and became plants containing the nutrients needed by animals and human beings. The evolution and growth that took place within the constraints of this solar income was a sustainable type of growth, since the amount of energy coming from the sun could be relied on to be nearly constant over time. Over the last two centuries, however, the growth of industry and agriculture has increasingly taken place beyond the constraints of solar income and now depends heavily upon fossil fuel capital. Many of the technological successes of the industrial revolution have revolved around new ways to use fossil fuels as a mechanical labor substitute in tasks originally done by human beings and draft animals. The contemporary dilemma is that growth driven by fossil fuels is not indefinitely sustainable, since supplies of the most useful of these "fund" sources of energy are finite in relation to projected human demand for them, and there are increasingly severe environmental constraints on their use.

The history of the industrial revolution has been marked by a progression from coal to petroleum and eventually natural gas, not only to perform the tasks originally done within the constraints of current solar income but also to expand dramatically world industrial production. In fact, the world population is now so large that it would be impossible for the bulk of it to survive a sudden transition back to solar power. The events of the two energy crisis cycles serve as a reminder of the dangers involved in becoming more heavily dependent on these finite and geographically concentrated resources and related vulnerable global energy supply systems in a very insecure world.[1]

Besides heightening awareness of the energy predicament, the first two energy crises focused attention on the social psychology of the energy problem. During the onset of each crisis cycle, individuals, corporations, and nations scrambled to stockpile petroleum and petroleum products, resulting in major shortages and huge price increases. But after the first major oil shock, people quickly forgot the conserving behavior induced by the crisis; and many of the so-called experts boldly stated that such a crisis simply couldn't happen again.[2] Yet, in the second crisis it took only a 4 percent oil shortfall during one quarter of the year to induce another panic and double the per barrel price.[3]

Thus, if history is a guide, the human ability to learn from recent experience seems limited, and the ability to plan for a possibly more precarious energy future seems almost nonexistent. Shortly after the first crisis, the U.S. and world economies were on the mend. The oil crisis was considered by many optimists to be over, and energy consumption was once again on the increase. In the United States total energy consumption, which had bottomed out at seventy quadrillion British thermal units (BTUs) in 1975, once again returned to the 1973 level of seventy-four quads and then leaped to seventy-nine quads in 1979. Despite the lessons that ought to have been learned earlier, petroleum imports leaped from 6.1 million barrels per day in 1974 to a peak of 8.8 million barrels per day—just under 50 percent of consumption—in 1977.[4] And suddenly it all happened once again with the overthrow of the Shah of Iran in January, 1979. Even though the global petroleum shortage was extremely slight in 1979, oil companies, anticipating a deepening emergency, rushed to fill their storage tanks with crude and consumers rushed to fuel their automobiles with gasoline.

There is no way of predicting with precison when a third energy crisis will begin. Much depends on the future of political and military turmoil in the volatile Middle East. But U.S. energy consumption bottomed out at seventy quads in 1983, just as it did in 1975, and energy imports subsequently began to rise.[5] The battered world economy did not recover as rapidly in the 1980s as it did in the previous crisis cycle, and an energy surplus dominated the decade. But given the seeming political inability to learn and to plan, it is very likely that the previous energy boom and bust cycle will repeat itself and that, once again, a crisis will catch the energy importing countries unprepared.

▼ An Ecological Perspective

Two major energy crisis cycles have passed, and a glutted world petroleum market offers a respite during which to reflect on various aspects of the world energy problem. Constant conflict in the Middle East highlights an *immediate* national security dimension of the problem for the United States and other major industrialized oil importers. In general, the large consumers of oil have become increasingly dependent on distant producers, largely in the Middle East, and thus vulnerable to supply disruptions for political reasons. In the *medium term,* the price fluctuations associated with crisis cycles—as well as an inexorable increase in prices over time as petroleum and natural gas become more difficult to find—will continue to cause major economic adjustment problems. And in the *long term* a major ecological problem of resource depletion looms, which will necessitate a costly transition to alternative forms of energy production.

This long-term ecological aspect of the energy problem is an important point of departure, both because ecological factors condition the ongoing debate between oil producers and consumers over a just price for petroleum and because the location of the world's remaining petroleum and natural gas reserves shapes future economic and national security aspects of the problem. From an ecological perspective, the long-term problem for the United States, and eventually for the entire world, is that petroleum and natural gas, the preferred, clean-burning fuels, are in relatively limited supply. The United States is estimated to have less than 27 billion barrels in proved reserves, about one decade's worth of consumption relying solely on domestic supplies.[6] Comparable worldwide figures indicate 717 billion barrels in proved reserves, enough for about three decades.[7]

The energy forecasting business is filled with uncertainties of both supply and demand, and although geologists have a pretty good idea about general limits, there is no way that ultimate petroleum reserves can be known with precision. In nineteen major studies undertaken between 1942 and 1975, ultimately recoverable reserves of crude oil were estimated to be between 400 and 2290 billion barrels, with the most frequent projections centering around 2000 billion.[8] Although in recent years significantly more oil has been consumed than has been found in new discoveries, it is not out of the question that significant reserves may still be found in remote corners of the world.[9] In addition, it is impossible to project with precision the course of future global economic growth and thus the related demand for petroleum. Over the last decade, for example, world petroleum demand has varied between twenty and twenty-four billion barrels per year. Regardless of the optimism or pessimism of future projections, however, many geological characteristics of petroleum and natural gas make them unlikely candidates to be world energy backbones over more than a few decades.

Although the precise geological processes responsible for creating coal, petroleum, and natural gas are still not completely understood, most of the complex dynamics of fossil fuel creation are known in a general sense. Fossil fuels result from highly unusual geological events, which seem to have produced large quantities of them in relatively few locations. In this respect fossil fuels differ from other minerals, such as iron ore, which are more commonly found scattered across the earth's surface. It is generally accepted that almost all fossil fuels originated from solar energy. This once current solar income was converted into plant, and eventually animal, life through photosynthesis over more than three hundred million years. During that period, an exceedingly small percentage of dead plant and animal life was preserved through very rare occurrences that resulted in the creation of coal, petroleum, and natural gas. These rare processes worked at an excruciatingly slow pace; in relation to contemporary demand for the fossil fuels, creation of new supplies now can be considered insignificant.

Coal was the first fossil fuel to be extensively exploited during the spread of the industrial revolution. A dirty and bulky fuel, coal is found in copious quantities around the world. The six hundred billion tons of hard coal estimated to be in known reserves is adequate for more than two centuries of consumption at current rates of use.[10] Much coal represents the remains of trees once located in swampy areas near prehistoric seas. As the seas periodically expanded and contracted, trees would flourish, die, and fall into the swamps where they would be kept from rotting, or oxidizing, by water and deposition of sands and clays on top of them. Because of the geological conditions responsible for their formation, most coal fields have well-defined boundaries and are composed of seams found at various depths. The ancient trees, compressed by geological pressures over long periods of time, have become lignite, bituminous, or anthracite coal depending on the extent of the pressures. The seams represent various epochs in geological history, working from more ancient depositions to more contemporary events as one moves from depths to the surface.

Petroleum and natural gas are not nearly as abundant relative to present consumption as coal is, because the processes that created them were much more specialized. Petroleum and natural gas are associated, or found together, about 50 percent of the time in areas called *fields* and in accumulations called *pools*. Most fields are only a few square miles in size, and even the larger fields seldom exceed one hundred square miles. As in the case of coal, the material that is now petroleum and natural gas was once organic matter, tiny plants and animals that lived in prehistoric oceans and lakes. These small organisms died and precipitated to the lake and ocean beds, where sediments of clay and similar materials covered them. Layer upon layer of compacting shale pressured the organic matter, over long periods of time, into valuable petroleum and natural gas.[11]

Petroleum and natural gas are not always found where the creation process began, because they have been subjected to geological forces that have squeezed them from high-pressure to low-pressure areas. In general, these forces have moved them toward the surface, and a significant portion of these valuable resources has boiled away into the atmosphere. The quantities that remained in the ground usually migrated into porous source rocks, such as limestone or sandstone, and were kept in place by various kinds of impermeable rocks that created trapping structures. Thus, a rare combination of events, including preservation of microorganisms in sedimentary basins, tremendous pressure over time, movement into reservoir rocks such as sandstone or limestone, and protection by a nonpermeable trapping structure, must occur in sequence for petroleum and natural gas fields to persist. The joint occurrence of the necessary geological conditions and migrating liquid and gas hydrocarbons is exceedingly rare; this rarity explains why the world has so few significant petroleum- and gas-producing areas.[12]

Although there is a much greater worldwide abundance of coal, petroleum and natural gas burn much more cleanly and efficiently and are thus preferred fuels for home heating, transportation, and in many industrial processes. Without petroleum, the world's massive fleet of automobiles, trucks, trains, and planes would crawl to a halt. While the world will never actually run out of oil, commercially viable exploitation for heating, manufacturing, and transportation is expected to become problematic within thirty to fifty years. M. King Hubbert has done some of the most relevant work in projecting future oil reserves in the United States and the rest of the world. Assuming that the world economy continues to develop at much the rate that it has in the past, he has estimated that world petroleum production will peak just before the year 2000. According to his estimates, at that time the world will have consumed 80 percent of its reserves in a period of less than sixty years.[13]

Natural gas is in less limited supply but, since much of it is found associated with petroleum, the countries blessed with petroleum are usually also blessed with abundant natural gas. Because of transportation difficulties, however, the international gas market is not nearly as extensive as is the petroleum market. Gas is best transported by pipeline, and supplies are usually geographically contiguous with markets. Thus, the Soviet Union has built extremely long pipelines from Siberia to Western Europe in order to earn hard currency for gas exports, and a regional natural gas network has grown linking the United States with Canada and Mexico. But much of the natural gas associated with oil production in the Middle East is still burned because of the expense and difficulty of getting it to market via liquefied-natural-gas tankers. It is not easy to project accurately the pace of worldwide natural gas depletion because so much of it is found and used in the Soviet Union, but studies indicate that production in major consuming countries outside the Soviet Union will be at only two-thirds the level of the early 1970s by the year 2000.[14]

The ecological aspect of the energy problem is not simply one of dwindling future supplies of more desirable fossil fuels. Table 3-1 illustrates that current reserves of fossil fuels are distributed in a highly inegalitarian manner and that some major powers have significant fossil fuel–related security concerns. Some countries that now are big energy importers have never had significant reserves. Others, like the United States, have had large reserves but have already used most of nature's bounty. These figures also show that there has been remarkable stability in the petroleum

Table 3-1 World petroleum reserves.*

	1974	1978	1982	1986	1986 Production
Saudi Arabia	103	113	165	167	4.9
Kuwait	71	71	68	92	1.3
USSR	56	58	86	61	12.3
Venezuela	19	18	25	56	1.8
Mexico	3	28	48	55	2.5
Iraq	35	34	33	40	1.7
Iran	68	45	38	37	1.9
Abu Dhabi	40	30	35	35	1.1
United States	34	28	30	27	8.7
Libya	23	27	25	23	1.0
China	15	20	18	19	2.6
Nigeria	14	12	20	16	1.5
Norway	6	4	8	11	.8
Indonesia	12	8	10	8	1.4
Algeria	9	10	7	5	1.0
United Kingdom	8	10	7	5	2.5
World Total	569	567	681	717	57.0

*Reserves in billions of barrels; production in millions of barrels per day.

SOURCE: *World Oil.*

producer hierarchy over time. Saudi Arabia had by far the world's largest reserves in 1974 and has widened the gap between itself and other countries over the period. The sixteen countries with the largest reserves have remained almost the same, the only exceptions being the replacement of the Middle Eastern "neutral zone" and Canada by Norway and Mexico. Although a small amount of internal reshuffling has taken place among these petroleum elites, the only real newcomer to the group is Mexico, with its rags-to-riches rise from three to forty-eight billion barrels in reserves.

These data emphasize that oil fields are highly concentrated in those few regions that have experienced the rare combination of geological conditions required for formation of significant oil and gas reservoirs (see Figure 3-1). The most abundant supplies are in most cases located far from the markets for them. Saudi Arabian dominance of the present and, more important, the future international petroleum market is most obvious. The Saudis own one out of every four barrels of proved reserves, and they have increased reserves nearly 60 percent over the last decade. When the rest of OPEC (Organization of Petroleum Exporting Countries) reserves are added to those of the Saudis, two thirds of world oil is accounted for. Adding in Soviet reserves accounts for all but about one quarter of the total. Aside from Norway and Great Britain, with a meager fifteen billion barrels of reserves between them, no Western European countries have meaningful supplies. And Japan, one of the world's largest consumers of petroleum, has no domestic sources of any consequence.

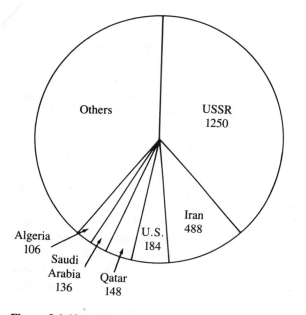

Figure 3-1 Natural gas reserves, in trillion cubic feet, 1986.
SOURCE: *World Oil.*

The United States faces a steady depletion of its petroleum and natural gas reserves. Although the present petroleum reserve of about 27 billion barrels is declining at about 4 percent per year, U.S. reserves were once very substantial. The problem is that since the first drilling for oil took place in 1859, the United States has produced more than 140 billion barrels of oil.[15] Thus, the U.S. was once almost as generously endowed with petroleum as Saudi Arabia, but most of this reserve has been used to support the heavy industrialization that took place in the United States over the last century. In recent years nearly all of the United States has been scoured in a search for new reserves, and little has been added. In 1984, before the impact of the oil price slump was felt, more than 75,000 oil wells were drilled in the United States in order to sustain production and explore for new reserves. In Saudi Arabia, by contrast, needs were met by drilling only 110 wells.[16] Thus, U.S. industry has had to turn to imports and alternative domestic energy sources to meet current and projected demands.

In summary, the ecology and geology of world oil and natural gas supplies has created political and economic vulnerabilities for the United States, Japan, and most of Western Europe. To continue traditional forms of industrial economic growth in these countries means becoming even more dependent on foreign energy supplies. Long and vulnerable supply lines now extend to these regions from the Middle East, where most of the world's future reserves will be found.

▼ The Economic Problem

Although there is still uncertainty over exactly when natural gas and petroleum will cease to be economically viable energy sources on a large scale, there is good reason to be concerned about the inevitable medium-term price increases and cyclical dislocations that can be expected as reserves dwindle and become more concentrated in the Middle East. Over the next few decades a tremendous amount of capital will be devoted to exploration for and development of new petroleum and natural gas reserves. As costs escalate, at some point it will make both economic and political sense to invest in alternative energy sources in order to soften the impact of the inevitable future energy crises.

Thus, the medium-term aspect of the energy problem is one of prudent political and economic management. Reasonably stable energy prices are essential for continued vigorous industrial growth around the world. The large price increases of the first two crisis cycles and the subsequent price collapses have had an extremely destabilizing effect on the global economy. Since energy is closely linked to almost all economic activity, rising energy prices send reverberations throughout the entire global economy as the initial shock wave is passed through in the form of higher prices for gasoline, electricity, industrial products, and even agricultural commodities. And although rapid price collapses are beneficial to oil-importing countries in the short term, they have a devastating impact on producers and exacerbate long-term problems because of cutbacks in exploratory activity. Price stability is thus in the interest of all parties in the world energy market. Petroleum and natural gas exporters prefer stability at very high prices and importers prefer stability at low prices, but sharp fluctuations in prices are destructive to all involved.

The first two cycles of energy crisis created massive trade deficits and rampant inflation in most of the industrial economies. These problems were followed by monetary and fiscal policy adjustments, leading to economic recession and eventually decreased demand for oil. The exporting countries fell heir to huge profits and a resulting dollar recycling problem as ports became clogged with ships bringing a deluge of new imports. Much of the mass of these new petrodollars was invested in gold or foreign bank accounts. The two cyclical surges in oil profits were followed by periods of substantial economic contraction, the second lasting through much of the 1980s. The recessions were responsible for dramatic declines in petroleum demand, and the subsequent downward spiral of oil prices created major debt repayment problems for economically weaker oil exporters such as Venezuela, Nigeria, Mexico, and Egypt.

If the world were a rational place, producers and consumers would come to an agreement on long-term pricing that would permit an orderly growth in consumption as well as an eventual transition to alternative energy sources. But negotiations aimed at price stabilization have been unsuccessful. U.S. officials in the Reagan administration took the position that market forces should be allowed to dictate prices in the hope that oil prices will remain depressed, whereas oil exporters are hesitant to come to an agreement on cheaper oil prices in the hope that the long-term trend toward higher prices will prevail. Oil-exporting countries point to a general agreement among experts that the most easily available petroleum and natural gas reserves are already being exploited. New discoveries are likely to be found at greater depths, in geographically remote or politically unstable areas of the world, or in coastal waters where drilling is very expensive. In each of these situations, higher costs will be associated with finding, drilling for, and transporting oil and natural gas to markets. As prices rise, enhanced recovery techniques can be used to get more petroleum and gas from existing fields, but these techniques also add considerably to the cost of petroleum delivered to refineries.

The greater difficulty involved in finding and drilling for future oil and gas is an important constraint on economic growth, because it represents the loss of a natural subsidy or source of surplus responsible for much of the growth of the industrial period.[17] When Colonel Edwin Drake drilled his first oil well in Titusville, Pennsylvania, in 1859, it cost him very little to strike "black gold." Only small amounts of energy or money had to be expended to produce petroleum found so near the surface, and most of the oil recovered represented a large net energy gain. Even though new technologies aid contemporary exploration and production, deeper drilling, drilling offshore, or drilling in remote locations costs more in energy and economic terms. Recently a consortium of oil companies abandoned a major exploratory operation off the Alaskan North Slope after investing more than a billion dollars in the effort. In the Baltimore Canyon area off the East Coast of the United States, exploratory activity by a number of companies was called to a halt after an investment of over three billion dollars in forty-six dry holes.[18] The energy and economic subsidy available in each barrel of new oil will continue to decline as the amount of energy and capital expended to find and get each additional barrel comes closer to the total energy value.

The United States, the first country to exploit the bulk of its petroleum reserves, is already experiencing the increased costs associated with petroleum depletion. During the period of peak prices in the early 1980s, despite major efforts by domestic oil companies to find additional supplies, U.S. reserves dropped from thirty-four to twenty-seven billion barrels. Before the price slump, approximately 75,000 wells were drilled using 350 million feet of pipe in the peak year of exploratory activity. This represented more than three quarters of all the wells drilled in the world in that year.[19] Despite all this investment, reserves declined slightly and production remained stagnant. Between 1978 and 1983, fifteen major U.S. companies reported domestic additions to reserves from drilling and enhanced recovery that averaged only 44 percent of their yearly production. Comparative data for 1978–80 and 1981–83 show that fourteen major U.S. oil companies saw the average number of barrels added from each one hundred dollars spent on drilling and improved recovery drop from 14.5 to 8.3.[20] Assuming that domestic oil companies want to maintain their normal pretax earnings of $6.70 per barrel, a world market price for oil below $18 per barrel would gradually wipe out the U.S. domestic oil industry.

Similar economic data are difficult to obtain for the world oil market because of the diversity of actors engaged in drilling and exploration and the variety of conditions they encounter. Given that the oil industry is government-controlled in many countries and that most companies keep cost information proprietary, it is difficult to come up with average costs for producing a barrel of crude around the world. But it is safe to say that the dynamics of exploration and recovery leading to higher costs in the United States are also being experienced in marginal producer countries. Internationally, for wells drilled outside the United States and the Communist countries, the yield per foot of well drilled has plummeted. In 1956, each foot drilled yielded 400 barrels. This figure peaked at 583 barrels per foot in 1971, and since then has been dropping dramatically, hitting 309 barrels per foot in 1983.[21]

It would seem difficult to reconcile this projected upward trend in per barrel costs for future oil with the slump in oil prices in the late 1980s. But the dilemma is more apparent than real. Projections of increases in future oil prices assume at least modest growth in demand for the product. In the early 1980s the global economy was in virtually a no-growth situation for more than three years. This dampened demand for oil just at the time that supplies were enhanced by new producers such as Mexico and China. The oil glut will undoubtedly be temporary and persist only as long as new marginal suppliers are able to stay in the market and no rapid global economic expansion takes place.

▼ Security Concerns

The existing and likely future distribution of world oil and gas reserves raises immediate political and security concerns for many of the world's industrialized countries. The United States, Japan, and most Western European countries are major importers of energy. Some countries, like Japan, have never had extensive energy reserves. In recent years, for example, Japan has produced a meager ten thousand barrels of oil per day, less than 2 percent of daily consumption. Other countries like Germany went through significant quantities of coal as they industrialized but have subsequently discovered little petroleum or natural gas to sustain domestic industrial activity. Thus, as Table 3-2 reveals, a number of industrialized countries import significant portions of the total energy they now consume. Although the United States imports large quantities of petroleum, it is relatively more self-sufficient in total energy consumption than many other industrial countries.

The national security dimensions of the energy problem become clearer when industrial country petroleum import needs are matched with the concentration of reserves in only a handful of exporting countries. Saudi Arabia possesses nearly one quarter of the world's proven reserves, and the percentage has been growing. The other four major Middle Eastern OPEC members together have slightly in excess of one quarter of world reserves. Thus, these five countries will likely control more than one half of the world's future oil supply. If anything should happen to interrupt the flow of petroleum out of the Middle East for an extended period of time, the energy-deficient industrial countries would once again be in dire straits.

The Middle East is one of the politically least stable areas in the world. The Arab–Israeli dispute, which periodically leads to bloodletting, is unlikely to be resolved soon. Resurgent Islamic fundamentalists are creating havoc in a number of countries

Table 3-2 Energy self-sufficiency (total energy production divided by consumption).

	1975	1984	Consumption Per Capita 1984*
Japan	.09	.10	3800
Denmark	.01	.16	4521
Italy	.17	.16	3105
Belgium	.15	.19	4939
Finland	.08	.22	5002
Switzerland	.23	.25	3733
Austria	.40	.28	4007
France	.23	.29	3923
Sweden	.19	.38	4703
W. Germany	.51	.45	5564
Hungary	.59	.56	3790
E. Germany	.72	.76	7600
Rumania	1.00	.90	4558
United States	.87	.91	9577
Poland	1.12	1.04	4494
Netherlands	1.40	1.15	5854
Canada	1.17	1.26	9773
USSR	1.23	1.29	5977
Australia	1.25	1.53	6128
Norway	1.15	3.83	6570

*Kilograms of coal equivalent.

SOURCE: United Nations, *Statistical Yearbook.*

and have targeted the governments of Saudi Arabia and Kuwait for destruction. Iran and Iraq, two key OPEC members, fought a war of attrition for years that periodically spilled over into the Persian Gulf. The proximity of the Soviet Union to the area raises the possibility of Russian mischief.

The industrialized Western countries have already had two experiences to help prepare them for future supply disruptions. The first occurred in late 1973, when the Organization of Arab Petroleum Exporting Countries (OAPEC) used the oil weapon selectively to embargo Western nations deemed sympathetic to the Israeli cause. Although the embargo was only a partial success, it demonstrated how easily oil supplies might be disrupted by a full OPEC embargo. The use of the oil weapon at that time shook up the existing political and economic order. Politically, leaders in oil-exporting countries who had been largely ignored in international political circles suddenly were courted by diplomats from importing countries. Support for Israel softened among big oil importers, such as Japan, as maintaining access to Saudi oil reserves became a pressing issue. Perhaps of greater importance, industrial countries proved more than willing to offset higher oil prices by flooding the Middle East with arms. Economically, the embargo led to OPEC ratification of a new price structure, which represented a fourfold increase in oil prices. It has been estimated that in the first three years after the embargo oil importing countries paid an additional $225 billion for OPEC oil and lost $600 billion in economic productivity as a result of the ensuing recession.[22] The gravest effects of the recession were felt in 1974 and 1975, when growth rates in industrial countries plummeted from a pre-embargo average of 5.4 percent in real GNP to −1.4 percent in 1975. The non-oil-exporting less developed countries experienced a similar drop, from 6.6 percent real GNP to 1.9 percent.[23]

The second major oil supply disruption started in 1979, mostly as a result of the Iranian revolution. Although the actual disruption was small and short-lived, the psychology of scarcity ruled the spot market, and prices for oil doubled, an event that was again duly ratified by OPEC after the fact. The global recession that followed on the heels of this disruption proved to be more lasting. In the industrial countries there was very little growth in real gross domestic product from 1980 through 1982, with only very modest increases in 1983. Real gross domestic product growth in the developing countries dropped below 1 percent in 1983.[24]

American political leaders have had difficulty coming to grips with these new national security realities. Richard Nixon's exclusionist solution to increased U.S. reliance on imported oil was "Project Independence," a high-technology effort that was supposed to make the United States free of energy dependence by developing alternatives to oil. After using up considerable capital the project yielded few significant results. At various times over the last decade, U.S. leaders have advocated military intervention in the Middle East as a solution if access to "American" oil is threatened by turmoil in the region. The now defunct Rapid Deployment Force was created for just such a contingency. Although the United States did create a significant strategic petroleum reserve, the Reagan administration attempted to dismantle the Department of Energy and came up with no real energy emergency plans for the country.

Leaders of other OECD countries have taken the security implications of the energy problem more seriously, since they are in general not nearly so energy self-sufficient as the United States. Although Norway, the United Kingdom, and the Netherlands produce significant quantities of petroleum and natural gas from the North Sea, Japan and the rest of Western Europe would be much more affected by another cutoff of Middle Eastern supplies than would the United States. This difference in vulnerability would likely create friction among the allies in a new crisis. To help remedy the problem, the OECD countries created the International Energy Agency to oversee sharing arrangements should a dramatic shortfall of petroleum occur, but it is doubtful how effective such arrangements would be during a serious confrontation with oil exporters.

▼ Energy Problems East and West

There has been a very close relationship between the spread of industrial growth and increased energy consumption. The definitions of progress at the core of the dominant social paradigm of industrial societies stress energy-intensive automation featuring substitution of fossil fuels for human and animal labor. Highly industrialized countries are thus characterized by elevated levels of per capita energy consumption in comparison with their less developed counterparts. But not all industrial countries are equally judicious in their use of energy. For example, the United States historically has been a profligate energy consumer, in large part because of what seemed in the past to be boundless energy reserves. Sweden, by contrast, has supported a similar standard of living with much less energy consumption.[25]

The United States became a major importer of petroleum in the late 1960s, and net imports peaked at about 8.6 million barrels per day in 1977. This represented about 47 percent of U.S. consumption.[26] During the subsequent recession, U.S. imports dropped considerably, but lower oil prices again led to increased imports in the late 1980s. The United States has considerable coal reserves and significant supplies of natural gas that lessen the overall dependence problem. The United States used to be dependent on oil imported from the Middle East. But the data in Table 3-3 indicate that American oil companies have diversified their sources of petroleum and now concentrate on the western hemisphere. In 1977, for example, about two thirds of all oil imports came from OPEC countries. In recent years this figure has been cut in half. At present, Venezuela, Mexico, Canada, Saudi Arabia, and Great Britain provide the United States with most of its imported oil.

In contrast with the energy-importing industrialized market economies, the Soviet Union has emerged as the world's largest producer of petroleum. Soviet production and reserves have been the subject of major controversies in the U.S. intelligence community. In 1977 the Central Intelligence Agency issued a report predicting that Soviet oil production would peak and decline in the early 1980s and that the Soviet Bloc would become a significant importer of oil shortly thereafter.[27] In fact, the Soviet Union doesn't face the same type of depletion constraints that are draining the American petroleum industry. The eighty-five-billion-barrel Soviet petroleum reserve

Table 3-3 U.S. petroleum imports, in thousands of barrels per day.

	1973	1977	1982	1987
Canada	1325	517	482	837
Venezuela	1135	690	412	768
Mexico	16	179	685	645
Saudi Arabia	486	1380	552	747
Nigeria	459	1143	514	530
United Kingdom	15	126	456	368
Indonesia	213	541	248	277
Algeria	136	559	170	284
OPEC	2993	6193	2146	2994
Non-OPEC	3263	2614	2968	3547

SOURCE: U.S. Department of Energy, *Monthly Energy Review.*

estimate is probably conservative, since vast tracts of the Soviet Union have been inadequately explored. In addition, Soviet natural gas and coal reserves are the largest in the world.

Soviet petroleum production has been sustained at about 12.5 million barrels per day since 1983, a figure that is about 3.8 million barrels per day higher than U.S. production. Soviet domestic demand does not require all of this production, and about 2 to 3 million barrels of petroleum per day are available for export to Eastern and Western Europe.[28] While petroleum production has stabilized, natural gas production capacity has been steadily growing, and the Soviet Union is developing significant export capability to offset hard currency losses should petroleum exports begin to taper off. When all forms of energy are considered, the Soviet Union is an energy giant, capable of helping Western Europe and Japan meet their future energy needs—a fact that is a source of much discomfort to U.S. policy makers who worry about the strategic consequences of this resource treasury.

Although the Soviet Union has extensive reserves, its oil industry is not without problems. The supergiant Samotlar field in western Siberia, which has accounted for about one quarter of all Soviet oil production, is beginning to decline, and other fields to be developed are in much more inaccessible regions of Siberia. Natural gas reserves are also located in cold and remote reaches of the country, and their development poses similar problems. The Soviet task is to gain access to new technology and capital essential to the development of new wells and long pipelines leading back to the western parts of the country, where most Soviet industry is located.[29]

From an inclusionist perspective, the energy vacuum in Western Europe combined with the abundance of Soviet reserves would seem to provide the basis for long-term cooperation. Western European banks and industries have the capital and technology the Soviets need to develop new oil and gas reserves. The Soviet Union has abundant natural gas reserves that could help Western Europe diversify away from tenuous Middle Eastern supplies. Thus, in the early 1980s the Soviet Union completed a thirty-six-hundred-mile pipeline connecting the Urengoi and Yamburg natural gas fields in Siberia with markets in Western Europe. The United States opposed European participation in the project and went to great lengths to discourage the transfer of pipeline technology to the project.

The pipeline dispute underlined some fundamental differences between U.S. and Western European perceptions of Soviet intentions. Western European leaders see projects such as the pipeline as a way of fostering greater East–West cooperation. They hope for a continuing thaw in East–West relations in order to take advantage of a growing Soviet market for Western technology. They also see the Soviet Union as a more reliable supplier of energy than many Middle Eastern countries. Many American politicians see such cooperation as a threat to Western European security, pointing out that fuel dependency could be exploited during times of international tension. Additionally, certain hard-liners have argued that the currency earned by the Soviets could be used to modernize Soviet military forces. But U.S. efforts to halt construction of the Soviet natural gas pipeline bore little fruit, and deliveries to Western Europe commenced in the mid 1980s.[30]

▼ The World Oil Market

A continuing management problem exists in the world energy market because the world's economies are now tied tightly together in an integrated energy production-consumption network. The largest portion of energy exchange takes place in the petroleum market and petroleum now accounts for nearly half of all measured energy consumption in the non-Communist world. Although natural gas and coal each account for nearly one fifth, only a small percentage of the production of each actually enters into international trade. Thus, it is petroleum that moves in large quantities across national borders, and petroleum will continue to dominate the energy market until well after the year 2000.[31]

The world oil market has a long and somewhat less than noble history, and many of its peculiar characteristics have persisted over the years. The oil industry got its start in 1859 when Edwin Drake drilled the first successful well near Titusville, Pennsylvania. During the initial decade of production many of the problems that would bedevil the industry over the years were already apparent. Initial high prices for the precious fluid gave way to cheap oil as producers rushed to the Pennsylvania oil fields in order to cash in on the bonanza. Oil that was priced at twenty dollars a barrel in the first year after the Titusville discovery was soon selling for less than ten cents. In subsequent years there were pitched battles among producers, all of whom were struggling to produce more oil, and between producers and owners of downstream facilities, the transportation, refining, and marketing networks. It was under these circumstances that John D. Rockefeller took an active role in organizing the market and restricting competition. The results of his shadowy activities were felt for more than a century in the world petroleum market.[32]

Rockefeller first got into the oil business in a very unusual way. He didn't buy land, drill for oil, or do anything else the early wildcatters did. Instead, after assessing the situation, he decided to buy a refining operation in Cleveland. Operating in secret and often using ruthless tactics, Rockefeller was able to profit through control of the downstream flow of petroleum after it left the wellhead. He cut deals with railroads, and built refineries and retail operations as the industry moved west. By 1883 Rockefeller had consolidated his hold on a fledgling nationwide petroleum industry in what was then known as the Standard Oil Trust. He bought oil fields to fully integrate his company from the wellhead to the customer, and then he moved into the international market. By 1885 over 70 percent of his business was done overseas.[33]

The size of the Rockefeller empire and the arrogance of its officers inevitably led to a public outcry, which found political expression in the Sherman Antitrust Act of 1890 prohibiting all business combinations in restraint of trade. Although it took years for the government to develop the courage to attack the well-entrenched Rockefeller empire, in May 1911 the U.S. Supreme Court declared that Standard Oil was guilty of violating the antitrust laws and ordered it to divest itself of all its subsidiaries, thus dividing the Rockefeller empire into thirty-eight separate chunks. Although this dissolution ended the first round of monopoly in the world oil industry, three of the progeny—Exxon, Mobil, and Socal (Chevron)—quickly became dominant forces in their own right and joined with four other companies to become the Seven Sisters, the multinational companies that subsequently developed a nearly complete monopoly of the world oil market.

By the end of World War I, the chaos that followed the breakup of the Standard empire gave way to a market dominated by the Seven Sisters, made up of the three Standard offspring and two American and two European oil companies that had emerged with the clout necessary to dominate the industry. In the United States, Gulf and Texaco established a beachhead in Texas and became major forces in the market. Gulf was financed by the Mellons out of Pittsburgh and grew courtesy of the great Spindletop discovery in Texas and similar finds in Oklahoma. Texaco also developed in Texas, marketing cheap oil from Spindletop and Sour Lake on a regional basis. The five American companies were joined by Royal Dutch Shell, a British-Dutch company that made its mark in the Far East, and British Petroleum, which owed its origins to oil from Burma and the Middle East, to form a cartel that controlled the international market for decades.

There was no love lost among the heads of these major oil companies, and considerable intrigue took place among them after the Standard demise. But by 1928 the Seven Sisters had knit together an organization that dominated the international market. This was accomplished partly through meetings held in Ostend, Belgium, in July 1928 at which oil concessions in the former Ottoman Empire were divided up among four of the Seven Sisters. In the so-called Red Line agreement, a line was drawn around the former Ottoman territories, which included oil-rich areas in Saudi Arabia and Iraq, and American and British companies agreed to operate through a joint venture, the Iraq Petroleum Company, to control production in and limit access to what were to become the great oil fields of the Middle East.

In the same year the big three contenders for dominance in world oil markets—Exxon, Shell, and British Petroleum—undertook a series of secret meetings intended to dampen increasingly hostile competition among them. Culminating in a grouse shoot at Achnacarry Castle in Scotland, the meetings defused a potentially explosive confrontation among the three. Real price wars had broken out in various parts of the international market largely because of a glut of oil. New production was coming out of Russia, Venezuela, Mexico, and Iraq. The Seven Sisters, in an atmosphere of distrust, began flooding each other's markets with cheap oil. Thus, leaders of the big three settled in for two weeks of meetings, dining, hunting, and fishing, which culminated in an agreement that was not fully disclosed until 1952.

The Achnacarry Agreement laid out the key principles by which the three multinational companies, and later the other four sisters and eleven smaller American companies, agreed to control the international market and deal with the production glut. Each company agreed to maintain its current volume of production and accept a similar proportion of any future production gains. New facilities were to be added only in response to increased public demand for petroleum. World prices were to be determined by a "Gulf Plus" pricing formula. This formula, developed to protect increasingly expensive American oil in a marketplace experiencing cheaper Middle Eastern competition, priced oil delivered anywhere in the world at the U.S. price plus freight charges from the Gulf of Mexico. Thus, oil being shipped from the Middle East to Western Europe would be billed at the Texas price with freight charges added as if the cargo originated in the Gulf of Mexico. In addition to providing a price-fixing formula, the agreement gave rise to additional profits from phantom freight— arrangements made by the Seven Sisters to swap production to decrease shipping costs, while pocketing the profits for trips from the United States that were never actually made.[34]

With small refinements, the Achnacarry Agreement put in place a market structure that persisted into the late 1950s. Although the entire world market could never be completely controlled, because of significant oil production coming out of Russia and the emergence of several smaller petroleum companies, the Seven Sisters controlled enough of the world's downstream facilities effectively to control prices. By the mid 1950s the Seven Sisters, whose vertically integrated operations included exploration, drilling, transportation, refining, and merchandising, controlled about 90 percent of the downstream facilities outside the Soviet Union. But toward the end of World War II the activities of various of the Seven Sisters came under scrutiny, particularly when it was discovered that oil was being sold by the same companies to France at 95 cents per barrel, to Uruguay at $1.00 per barrel, and to the United States Navy for $1.23 per barrel. But the near monopoly of the oil cartel remained largely in place until new competitive forces began to chip away at its foundation in the 1960s.

▼ The OPEC Decade

The less developed oil-exporting countries were relatively powerless in their dealings with the Seven Sisters during the first half of the twentieth century. Most of them were not even really countries but were passed as protectorates or colonies from one country to another as a result of political and military events. Even though formal independence came to many of them as part of the agreements carving up the Ottoman Empire after World War I or as part of the more general independence movement following World War II, they were in no position to assume a strong posture in negotiation with the Seven Sisters, which were backed by the political and military clout of their powerful governments. Mexico tried to deal resolutely with foreign oil companies in 1938, and Iran tried similar tactics under the socialist Mossadegh regime in 1951, but both countries were severely disciplined for their efforts. In each case the British broke diplomatic relations with the offending country, and foreign markets for its oil dried up. In the case of Iran, Mossadegh was replaced in a counter-revolution aided by outside parties.[35]

Ferment that would eventually result in the destruction of the Seven Sister monopoly and the rise of the oil-exporting countries as the organizing force in the market began slowly in the 1950s. On the heels of World War II, competition once again began to heat up in the petroleum market. Small independent oil companies that had been simply tolerated in earlier decades began to move more boldly to exploit new oil reserves outside the domain carved out by the Red Line agreement. These independents—among them Occidental, Phillips, Sun, and Atlantic Richfield—were able to get a toehold in the markets dominated by larger companies by buying and selling petroleum as cheaply as possible. The so-called majors were forced to confront the challenge of the independents by adjusting their own prices downward. This vigorous price competition resulted in the gas wars of the late 1950s and early 1960s. In order to compensate, the majors, who weren't about to absorb the full impact of the lower prices, forced producing countries to take a substantial cut in royalties derived from the posted price of a barrel of crude. In 1948 the posted price of a barrel of crude was $2.17. By 1960 the majors slashed this to $1.80, and subsequent cuts brought the actual selling price of a barrel of crude down to as little as $1.30 by the early 1970s.[36]

The emergence of the Soviet Union as a major oil exporter also served to loosen up the market and drive oil prices down. Sitting on massive reserves of petroleum, the Soviets needed to export large quantities of petroleum to get hard currency for rebuilding their war-torn economy. The Soviets sold and bartered petroleum at prices designed to capture the market from established exporters. And the independents were only too happy to buy this cheap oil, which began bubbling into storage tanks throughout the world.

National oil companies also began to play a larger role in buying and distributing petroleum. For example, the Ente Nazionali Idrocarburi (ENI) was established to obtain cheap oil for Italy, a country almost devoid of petroleum reserves. This enterprise involved bulk purchases from producer countries and further undercut the strong hold of the Seven Sisters on the world market.

The most important result of these developments was not only cheaper gasoline and oil for consumers but also extreme discontent among oil exporters, who absorbed much of the downward pressure on prices. Venezuela was among those most affected.

Higher costs of production for Venezuelan crudes, long distances from European markets, and restrictions on U.S. oil imports combined to drive Venezuela nearly out of the oil business. Thus, in response to price cuts that made future production in Venezuela marginal, Venezuelan leaders undertook the task of forging an organization of producer states that they hoped would be able to counter the power of multinational oil companies. OPEC was officially founded in 1960 with Venezuela, Saudi Arabia, Kuwait, Iraq, and Iran as charter members.[37]

During the first decade of its existence OPEC had relatively little impact where the vital interests of the oil companies were concerned and acted very much like a trade organization. The oil companies controlled downstream transportation, refineries, and retail outlets. They remained in a strong position from which they could discipline unruly leaders through selective buying of crude. Furthermore, during this period new oil discoveries in various parts of the world added to supply, thus giving the oil companies an opportunity to pit petroleum exporters against one another in a scramble to sell as much oil as possible. Even within OPEC, a series of meetings was required to formulate an action plan. The conservative Saudis, their neighbors in Kuwait, and the pro-U.S. shah in Iran were not about to take any dramatic actions that Venezuela or Iraq might suggest.

OPEC became a powerful force in organizing the world petroleum market largely because of a confluence of events over which the organization had very little control. The closing of the Suez Canal in 1967 as a result of Arab–Israeli hostilities marked a watershed in relations between exporting countries and the oil companies. Although the closing resulted in no major oil crisis, it did reveal the vulnerability of Western European importers, who were substantially increasing their dependence on Middle Eastern oil at that time. The canal closing led to only slight increases in prices as new supertankers were developed to carry oil around the tip of Africa to European markets. But the temporary dislocations gave the OPEC countries a slight taste of power, which was a precursor of things to come.

The market that developed subsequent to the canal closing was tightened in 1969 by disruption of the trans-Arabian pipeline in Syria, resulting in a reduction of five hundred thousand barrels per day in the oil supplied to the Mediterranean. This was followed by output cutbacks in Libya of eight hundred thousand barrels per day for "conservation" reasons. By 1971 there was little spare production capacity left in the United States, although about four million barrels per day of excess capacity remained in the Persian Gulf. In this new, semitight market, with almost all excess capacity in OPEC hands, the organization was able to assume a much tougher stance in negotiating with the oil companies.[38] Libya was particularly successful in early negotiations with Occidental Petroleum. Realizing that Occidental had few other options in a tight market, the radical Libyan regime was able to exact a thirty cent per barrel increase in the posted price of crude as well as other benefits, thereby setting a pattern for subsequent ratcheting upward of prices in negotiations between other producers and the major oil companies.

Three factors besides tightening markets have been identified as ultimately leading to the OPEC price revolution. Inflation ran rampant through the industrial world in the late 1960s and early 1970s, turning the terms of trade sharply against the oil exporters. While prices for manufactured products were rapidly rising, the price of oil remained constant or declined, thus diminishing the purchasing power of the oil-exporting countries. Second, a rising tide of Arab nationalism upset established political regimes in the area and led to demands for a different type of relationship with the oil companies. Finally, the continuing Arab–Israeli dispute added a political dimension to the world oil market.[39] All these factors combined to give OPEC more leverage in its relations with the oil companies. In a meeting between exporting countries and oil companies held in Teheran in February 1971, the producers were able to demand an increase in posted prices of thirty cents per barrel, and the majors capitulated. Thus, at the beginning of the Arab–Israeli war of October 1973, the price of a barrel of Saudi marker crude had edged up to $2.55.

But the really significant increases in oil prices during the OPEC decade were catalyzed by the two major oil crises that originated outside the formal OPEC organization. The first crisis was triggered by the 1973 hostilities in which the ten-member Organization of Arab Petroleum Exporting Countries declared a selective embargo against four different groups of countries. These exporting countries agreed to cut back oil production significantly and to deny oil to the United States, the Netherlands, Portugal, and South Africa. Remaining countries were placed in one of three other categories and received oil in proportion to their perceived sympathy for the Arab cause.

The embargo and production cuts were only partially effective. The non-Arab members of OPEC continued to produce at capacity, thus easing the impact of the Middle Eastern cuts. There was also considerable transshipping of oil from nonembargoed to embargoed destinations. Moreover, the embargo really lasted only five months. In the two months when it was at its peak, November and December 1973, the embargo removed only about 10 percent of the non-Communist world oil supply from the market. But the psychological and economic effects of the disruption were considerable. Countries and companies became locked in a struggle to maintain access to available oil supplies. Spot market prices in Rotterdam rose dramatically, and in this panic atmosphere, OPEC members had little trouble coming to agreement on higher prices. By the end of 1973, the price of a marker barrel of crude went to $3.60; by the end of 1974, it had jumped to $10.46. Once having ratified these dramatically higher prices and having felt no repercussions from oil companies or industrial importers, OPEC was in a solid position to supplant the Seven Sisters in controlling the market.

The rapid increase in oil prices had a devastating effect on the international economy and resulted in a global recession that dampened demand for oil in both the developing and industrialized countries. In the OECD countries consumption of petroleum dropped from an average of 39.6 million barrels per day in 1973 to an average of 36.1 million barrels per day in 1975. In the United States, the drop was from 17.3 million barrels per day in 1973 to 16.3 million barrels per day in 1975.[40] But the recession resulted in price stability in the oil marketplace. The Saudi marker barrel edged only very slowly upward in price, reaching $12.09 by January 1977.

The lessons of 1973–1974 were quickly forgotten as the international economy began to pick up steam again in 1976. Oil consumption in the OECD countries rose rapidly in 1976 and exceeded the 1973 average by about 750,000 barrels per day in 1977.[41] A certain "happy days are here again" euphoria developed in the importing countries, and many of the experts argued that there would never be another oil crisis. Petroleum consumption rose significantly and peaked in the OECD countries at 41.6 million barrels per day in 1979. In the United States a consumption peak of 18.9 million barrels per day was reached in 1978. OPEC oil production increased apace and peaked at 30.9 million barrels per day in 1979.[42]

The euphoria and price stability lasted until September 1978, when political events in Iran set the second energy crisis in motion. Anti-shah violence erupted in the cities and oil fields in Iran during the last quarter of 1978, resulting in a slowdown of production and a total halt in export activity by the end of the year. By March 1979, the new revolutionary government had resumed exports, but the panic had already had an impact on price. Similar to the first energy crisis, the initial signs of shortage pitted companies and countries against one another in a rush to get access to supplies of crude. Even though the actual shortage for the whole of 1979 was less than 2 percent of demand, panic drove prices up and OPEC was once again able to take advantage of the situation by raising its posted prices. By November 1979, the marker barrel price had reached twenty-four dollars and many countries were selling oil at a significant premium over OPEC official prices. The price for a marker barrel of Saudi crude hit its all-time high of thirty-four dollars per barrel in October 1981.

In only a decade OPEC had been transformed from a relatively powerless trade association into a powerful political and economic force capable of wresting control of the world oil industry from the Seven Sisters and their associates. But this meteoric rise to power covered over many serious differences among OPEC members that could subsequently destroy the cartel. It is extremely doubtful that OPEC could have had nearly the success it enjoyed in raising oil prices had it not been for the military and political events that took place in the unstable Middle East and the panic reactions that followed in the oil industry and the importing countries. Once the higher prices were firmly established, however, they provided a cement to hold this disparate group together, until the OPEC decade came to a screeching halt when oil prices began plunging in the mid 1980s.

▼ The End of the OPEC Era?

The world energy market changed dramatically in the 1980s. Repeating the cycle of events that followed the first oil shock, a global economic recession reduced significantly the demand for energy in general and for petroleum in particular. The cutback in oil consumption was deeper and more enduring in the first half of the 1980s than it had been a decade earlier. Collective demand for petroleum in the OECD countries peaked at just under 41.6 million barrels per day in 1979 and began to tumble rapidly. Over the next four years demand fell steadily until it bottomed out in 1983 at 33.9 million barrels per day.[43]

OPEC unity has been sorely tested by lagging petroleum demand and by the new petroleum contributions of marginal non-OPEC suppliers. For lack of markets, OPEC production plunged from an all-time high of 31.3 million barrels per day in 1977 to a low of about 16.1 million barrels per day in 1985.[44] Because of the risks involved in becoming too dependent on OPEC oil—particularly oil coming from the Persian Gulf—many importing countries buy OPEC oil only as a last resort, preferring first to use energy from fuels other than petroleum and then oil from non-OPEC petroleum-exporting countries. Thus, OPEC bears the brunt of significant swings in demand, rapid declines in sales when importing economies are weak and big increases in sales when these economies heat up.[45]

Maintaining OPEC cohesion is a classic public or collective good problem.[46] Each member has an obvious interest in the persistence of the organization. If the cartel should fall apart, all would suffer since oil prices would plummet rapidly. But each member also tries to exert as much political and economic leverage as possible within and outside the organization in order to capture a larger share of the oil market. The politics of adjustment to low demand and prices are obvious at each of the biennial and special OPEC meetings, as the OPEC countries focus on the price of a marker barrel of crude oil, differentials for other types of crude, the production quota to be assigned to each member, and the cheating that takes place as certain members circumvent their production quotas.

Not all of the world's oil is equally attractive to customers, and these quality differences give rise to considerable maneuvering within OPEC. Some oil is very heavy and lends itself readily to the production of asphalt, lubricants, and heating oil. Lighter oil is more sought after because it can be used to create larger fractions of economically desirable products such as gasoline or jet fuel. In order to keep petroleum produced by each member equally attractive in the marketplace, very early in its deliberations OPEC established a system of differentials in pricing based on specific gravity, purity, and distance to typical markets. The marker barrel of crude oil—the base on which the differentials have been computed—was a hypothetical barrel of Saudi Arabian light oil delivered to the Persian Gulf. Poorer quality crudes, such as the extremely heavy Boscan coming out of Venezuela, sell for several dollars less than the marker price, whereas lighter oils, such as Bonny Light from Nigeria, may sell for more than a dollar more than the marker barrel. To keep various kinds of oil competitive, these differentials have been adjusted over time to compensate for shifts in the marketplace and technological changes in refining processes. The adjustment process is a difficult one and a source of political friction within OPEC, since an adjustment of a nickel or dime per barrel can mean the gain or loss of millions of dollars.

There is also a continuing dispute among OPEC members over the price of a marker barrel. Price hawks press for raising prices in order to obtain the largest amount of revenue for limited reserves, while the doves are more worried about maintaining long-term growth in world demand. During the OPEC decade, the issue was to decide on a cap for prices. In the mid 1980s, however, market pressures drove the marker crude down nearly 50 percent, and price hawks such as Algeria, Libya, and Iran have only grudgingly adjusted to these "temporary" market realities and continue to press for much higher prices.

In 1982, market deterioration forced OPEC to establish member production ceilings for the first time. In previous years the marker barrel price combined with differentials was adequate to establish parity among OPEC members. But the extreme drop in world oil demand necessitated the painful allocation of a 17.5-million-barrel-per-day total quota for the organization. As the market continued to worsen, a temporary emergency quota of 16 million barrels per day was apportioned among the members in November 1984. The quota has remained within this range since then, but cheating has been widespread, since it is difficult for outside agencies to monitor national production.

The viability of OPEC over the next few years depends on the interplay of three sets of factors: the ability of very diverse countries within OPEC to settle their disputes within the constraints of a depressed oil market, the extent of future industrial growth and thus of demand for petroleum, and the role played by non-OPEC producers in supplying the market. OPEC is composed of countries with remarkably different cultural, economic, demographic, social, and political histories and prospects. They range from very wealthy countries such as Saudi Arabia and Kuwait on the one hand to very poor developing countries such as Nigeria or Indonesia on the other. Their petroleum reserves range from the 167-billion-barrel Saudi supply to the minuscule reserves of tiny Gabon (see Table 3-4). Their political systems range from representative democracies to military dictatorships and conservative theocracies. Yet the realization of large financial gains from cooperation has thus far provided a cement to hold the organization together.

OPEC members could be divided loosely into three factions that have common interests on many issues facing the organization. The first and most powerful group within the organization is that of the banker countries, so-called because during the OPEC decade they received so many petrodollars that they were forced to bank most of them. Led by Saudi Arabia, this group also includes Kuwait, the United Arab Emirates, and Qatar. These countries have very small populations and tremendous quantities of oil in the ground. Politically, they tend to be conservative, and in OPEC politics they are price moderates, concerned to nurture a long-term market for petroleum given their sizable reserves. Saudi Arabia has the production potential and the economic clout to dominate the banker countries and makes attempts to maintain order among the other OPEC countries. The Saudis for many years stabilized OPEC production and tightened the market through production cutbacks. But the Saudis can also increase production easily and flood the market if they choose to.

A second faction, consisting of Nigeria, Indonesia, Ecuador, Gabon, and Venezuela, is made up of developing countries. They are on the opposite side of the counter in the world banking system, being very dependent upon loans for continued development. Nigeria and Venezuela, in particular, have been hard hit by price and production declines in the 1980s and are numbered among the countries with significant international debt problems. These developing countries want both to pump oil at

Table 3-4 OPEC profile.

	1986 population (millions)	1986 oil reserves (billions of barrels)	1986 production per day (thousands of barrels)	Peak production (thousands of barrels per day)	Peak oil income (billions of dollars)	GNP per capita 1986 (dollars)
Saudi Arabia	11.5	167	4935	9532	109.4	6030
Kuwait	1.8	92	1344	2546	16.2	12708
Iran	46.6	37	1929	6022	18.2	1660
Iraq	16.0	40	1725	3477	24.8	2320
Abu Dhabi	1.4	35	1060	1999	19.6	14700
Libya	3.9	23	990	2175	22.1	5410
Venezuela	17.8	56	1791	3366	10.9	3030
Nigeria	105.4	16	1470	2302	19.1	590
Indonesia	168.4	8	1390	1686	18.2	430
Algeria	22.8	5	1000	1161	11.2	2510
Qatar	.3	3	239	570	5.4	23330
Ecuador	9.6	1	293	——	1.6	1120
Gabon	1.2	1	165	——	——	3300

SOURCES: Population Reference Bureau; *World Oil; The New York Times; Petroleum Intelligence Weekly;* Central Intelligence Agency, *Handbook of Economic Statistics 1987.*

maximum capacity and to maintain a high price for crude in order to accumulate capital for development. Nigeria is particularly disadvantaged by the fact that its Bonny Light crude competes directly with British North Sea Brent. As the British have stepped up production and cut prices, Nigeria has been forced into a dilemma: either cut its own prices and break ranks with OPEC, or watch oil sales deteriorate. Thus, Nigeria and the other less developed countries argue at OPEC meetings for special treatment on issues such as quotas, on the ground that their peculiar development situation requires a competitive edge.

The third OPEC faction could best be called the radicals. It is composed of Libya and Algeria in North Africa and Iran and Iraq in the Middle East. The fundamentalist Islamic regime in Iran is a thorn in the side of more conservative OPEC members, since it would like to topple them all from power. Iran and Iraq are perennially quarreling, which also disturbs OPEC unity. Libya and Algeria have been prominent for backing radical causes and have taken a lead role in the Group of 77, which demands a new international economic order. The radical group, with somewhat limited oil reserves, usually holds out for high oil prices based on their antipathy toward the oil-importing industrial countries.

Given this extremely diverse cast of characters, it is remarkable that OPEC has been able to persist in an atmosphere of lagging petroleum demand. Some new crisis arises at almost every OPEC meeting, whether inflicted from within or outside the organization. But the ultimate threat of a collapse in the oil market and thus the economic dismemberment of several members is a powerful incentive helping these diverse countries overcome their differences. It also should be noted that dissolution of the cartel and a resulting further worldwide decline in oil prices would be terribly disruptive to the international banking community, as well as the oil industry in the United States, and thus significant pressures for price stability also come from outside of OPEC.

The second factor shaping OPEC's future options is petroleum demand. Future energy demand cannot be projected adequately without knowing something about both the quantity and the quality of future economic growth patterns. If higher oil prices have indeed accelerated a post-industrial transformation of advanced industrial economies, those countries may never experience a large-scale recovery of oil demand. But there is no question that the global recession of the early 1980s was a primary factor in dampening petroleum demand and driving OPEC to a quota system. While the United States experienced an economic recovery in the mid 1980s, the then-strong dollar retarded economic recovery and related energy demand in other parts of the world. Petroleum is priced in dollars, and as the value of the dollar soared in relation to other currencies, the prices that others paid for oil also increased. Thus, a more general recovery outside the United States was stalled until the value of the dollar subsequently plunged.

There is also evidence that factors other than slow economic growth have begun to have an impact on lagging demand for petroleum. Primary among these is a structural transformation of the industrial economies away from the production of energy-intensive goods to the provision of less energy-intensive services. Should this transformation persist, white-collar service-oriented societies wouldn't require nearly as much energy per unit of GNP produced as would industrial, blue-collar equivalents. In addition, the twin oil shocks seem to have had at least some lasting impact on consumption patterns. Higher prices did result in energy conservation in homes, industry, and transportation. New energy-efficient technologies have at least temporarily lessened energy demand in the United States and other industrial countries.[47] In the United States, for example, the number of BTUs of energy needed to create a constant 1982 dollar of GNP steadily declined from 27,100 in 1973 to 20,100 in 1986.[48] Somewhat smaller increases in efficiency have also been noted in Western Europe and Japan, economies that were much more energy-efficient before the energy shocks. Whether these efficiency trends will continue in the face of much cheaper petroleum remains to be seen.

Finally, on the supply side new, marginal petroleum exporters outside OPEC have played a large role in adding to the oil glut (see Table 3-5). As a response to higher prices and out of economic necessity, many lesser exporters have developed their small reserves and have been exporting at levels near capacity. It is doubtful that they can continue to pump at this rate for many years or that other major sources of supply will emerge over the next decade. The Soviet Union has contributed about two million barrels of petroleum per day to Western markets, but numerous factors will cap Soviet petroleum production in the vicinity of 12.5 million barrels per day.[49] Mexico does have significant reserves and could expand production somewhat over the next decade. But Mexico will need more oil for domestic consumption in the next few years, and expansion of production will be inhibited by its large debt. Furthermore, Mexico has cooperated with OPEC in pricing and production and has little interest in breaking ranks in order to sell cheaper oil. The British are pumping significant quantities of oil from the North Sea, but these limited reserves will likely disappear over the next decade. Although there is some potential for new production

Table 3-5 Marginal oil producers.

	Production 1977*	Production 1986*	Reserves 1986**
China	1.81	2.62	18.5
Norway	.28	.84	11.1
United Kingdom	.74	2.49	5.3
Egypt	.41	.82	4.5
India	.21	.61	4.4
Malaysia	.18	.50	3.2
Brazil	.16	.59	2.4
Argentina	.43	.46	2.2
Angola	.17	.28	2.0

*Millions of barrels per day.
**Billions of barrels.

SOURCE: *World Oil.*

from China, the situation is much the same as it was a decade ago—a lot of promise, but limited production.[50] A number of other smaller suppliers became exporters of high-priced oil in response to the second energy crisis but have little potential to sustain exports in the 1990s.

One other factor could help turn the supply situation around in the early 1990s. During the early 1980s a significant part of the oil glut was created through the destocking activities of major oil companies. High interest rates and the prospect of cheaper future oil prices led many companies to reduce stocks to an extremely low level. Should the oil companies perceive a tightening market and the potential for higher prices in the 1990s, they might accelerate market tightening by rebuilding their stocks. Thus, any substantial economic recovery in the early 1990s could lead to a tightening oil market, restocking by the petroleum companies, and eventually a new petroleum decade for OPEC.

▼ A Third Energy Crisis?

The world has passed through two cycles of energy crisis during the last fifteen years. Each cycle was initiated by a rapid expansion of energy-intensive economic growth, which increased demand for oil and taxed the ability of the existing infrastructure to meet demands. In both cases political events precipitated a crisis by threatening to disrupt already tight markets. In both cases perceptions of crisis and hoarding behavior were a major part of the problem. Massive price increases caused major economic dislocations during both cycles, and a recession and lagging demand for petroleum followed. The major difference between the two cycles was that the second recession was deeper and longer than the first, and it coincided with new supplies of oil reaching the market from marginal producers, originally spurred on by higher prices.

How soon the world will experience a third energy crisis is really a function of how rapidly producers and consumers return to established patterns of doing things. There are many indications that the importing countries have already forgotten the lessons of the recent past. In the United States it has proved impossible to use the period of low petroleum prices to enact even a modest oil import tax as a cushion against future shocks. Economists and energy "experts" have gone back to writing optimistic articles very similar to those that were written in the late 1970s, right before the second oil shock. But the market dynamics that created the first two shocks still persist, and a third crisis cycle may well be under way.

On the supply side, the collapse of oil prices has given a short-term economic boost to the importers, but has had a devastating impact on exploration and development activities worldwide. Cheap oil makes it unprofitable to explore for and develop marginal fields. In the United States, for example, the 3970 rigs that were operating in 1981 were reduced to about 800 in late 1986 and have increased only slightly in number since then. The number of seismic crews in action fell from 681 to 155 during the same period.[51] Lower prices have led to many marginal wells being permanently shut in. Should the market tighten and prices rise in the early 1990s, there will be a producer-response delay of several years before new supplies reach the market. And all of this could take place at a time when current marginal exporter countries cease to be actors in the market. Low oil prices have also devastated alternative energy industries. The solar energy industry has nearly disappeared from view, and related research and development have fallen on hard times. Development of the nuclear power industry has pretty much halted, particularly in the United States, where cancellations of projects have exceeded new orders for many years.

Changes on the demand side are likely to be more complex and thus less easy to project with confidence. Lower prices are eroding voluntary energy-conservation activities and demand for larger automobiles is again rising. Whereas the structural transformation of industrial countries continues to keep increases in petroleum consumption there quite modest, consumption in the less developed countries is substantially increasing. The world stock market collapse and distortions in capital markets brought on by the U.S. trade deficit make the pattern of future global economic activity somewhat unclear. But regardless of the nature of future economic recovery, the dilemma for oil-importing countries in the 1990s is that a return to anything approaching traditional patterns of resource-intensive industrial growth makes a third energy crisis almost inevitable. It is ironic that the price that is likely to be paid for rapid future economic growth, should it occur, could be a third energy crisis cycle that would wipe out many previous economic gains with its associated recession.

▼ Notes

1. See Amory Lovins and L. Hunter Lovins, *Brittle Power: Energy Strategy for National Security* (Andover, Mass.: Brick House Publishing, 1982).
2. For example, see "The Coming Glut of Energy," *The Economist* (January 5, 1974); see also a precrisis view in M. A. Adelman, "Is the Oil Shortage Real?" *Foreign Policy* (Winter 1972–73).
3. Robert Lieber, "Energy Policy and National Security: Invisible Hand or Greedy Hand?" in Kenneth Oye et al., *Eagle Defiant: United States Foreign Policy in the 1980s* (Boston: Little, Brown, 1983), pp. 171–172.
4. Data from U.S. Department of Energy, *Monthly Energy Review* (July 1984), pp. 3, 43.
5. U.S. Department of Energy, *Monthly Energy Review* (September 1986), pp. 7, 47.
6. *World Oil* (August 1987), p. 25.
7. Ibid.
8. W.A.E.S., *Energy: Global Prospects 1985–2000* (New York: McGraw-Hill, 1977), p. 115.
9. See Exxon Corporation, *How Much Oil and Gas?* (May 1982), p. 14.
10. W.A.E.S., op. cit., pp. 170–171.
11. For a very understandable description of these processes see Exxon Corporation, *The Upstream: A Guide to Petroleum Exploration and Production.*

12. For more detail see Earl Cook, *Man, Energy, Society* (San Francisco: W. H. Freeman, 1976), pp. 75–87.
13. M. King Hubbert, "Energy Resources," in National Academy of Sciences, *Resources and Man* (San Francisco: W. H. Freeman, 1969).
14. W.A.E.S., op. cit., chap. 4.
15. Exxon Corporation, *How Much Oil and Gas?* p. 10.
16. *World Oil* (August 15, 1985), p. 29.
17. For a detailed discussion of the natural subsidy, see Earl Cook, op. cit., pp. 110–120.
18. Mark Potts, "Oil Hopes Dashed in Atlantic," *The Washington Post* (December 23, 1984).
19. *World Oil* (August 15, 1985), p. 29.
20. Dillard Spriggs, "Can U.S. Sustain Reserves at a Reasonable Cost?" *World Oil* (November 1984).
21. Figures from *World Oil* (January 1985), p. 29.
22. Estimated by *Business Week* (December 20, 1976), p. 45.
23. See Davis Bobrow et al., "Contrived Scarcity: The Short-Term Consequences of Expensive Oil," *International Studies Quarterly* (December 1977).
24. The World Bank, *World Development Report 1984* (New York: Oxford University Press, 1984), p. 11.
25. See Lee Schipper and Allan Lichtenberg, "Efficient Energy Use and Well-Being: The Swedish Example," *Science* (December 3, 1976). For a more general theoretical perspective see Glenn Hueckel, "A Historical Approach to Future Economic Growth," *Science* (March 14, 1975).
26. United States Department of Energy, *Monthly Energy Review* (September 1986), p. 41.
27. Central Intelligence Agency, *Prospects for Soviet Oil* (April 1977).
28. See *World Oil* (August 15, 1987), pp. 62ff.; *World Oil* (August 15, 1984), pp. 99ff.
29. See Office of Technology Assessment, *Technology and Soviet Energy Availability* (Washington, D.C.: U.S. Congress, 1981).
30. See Bruce Jentleson, *Pipeline Politics: The Complex Political Economy of East–West Energy Trade* (Ithaca, N.Y.: Cornell University Press, 1986), chaps. 6, 7.
31. "World Energy: Supply and Demand to the Year 2000," *Petroleum Economist* (November 1984), pp. 405–406.
32. For more detail see Anthony Sampson, *The Seven Sisters* (New York: Viking, 1975), chap. 2. See also Theodore Moran, "Managing an Oligopoly of Would-Be Sovereigns: The Dynamics of Joint-Control and Self-Control in the International Oil Industry Past, Present and Future," *International Organization* (Autumn 1987).
33. Ibid., p. 26.
34. Ibid., chap. 4.
35. See ibid., pp. 140–153.
36. Jahangir Amuzegar, "The Oil Story: Facts, Fiction and Fair Play," *Foreign Affairs* (July 1973).
37. For more detail see Ian Seymour, *OPEC: Instrument of Change* (New York: St. Martin's Press, 1981), chap. II.
38. Ibid., pp. 55–57.
39. Neil Jacoby, *Multinational Oil* (New York: Macmillan, 1974), pp. 258–259.
40. U.S. Department of Energy, *Monthly Energy Review* (September 1986), p. 113.
41. Ibid.
42. U.S. Department of Energy, *Monthly Energy Review* (December 1986), pp. 117, 119.
43. Ibid., p. 119.
44. Ibid., p. 117.
45. Bijan Mossaver-Rahmani, "The OPEC Multiplier," *Foreign Policy* (Fall 1983).

46. See Mancur Olson, *The Logic of Collective Action* (Cambridge, Mass.: Harvard University Press, 1965), pp. 14–16.
47. For an overview of these shifts see Joel Darmstadter et al., *Energy Today and Tomorrow* (Englewood Cliffs, N.J.: Prentice-Hall, 1983), chap. 2.
48. U.S. Department of Energy, *Monthly Energy Review* (December 1986), p. 16.
49. See Danilo Rigassi, "Gorbachev Spurs Oil Recovery," *World Oil* (August 15, 1987).
50. See Kim Woodard, "China's Energy Prospects," *Problems of Communism* (January–February 1980), p. 64.
51. Figures from U.S. Department of Energy, *Monthly Energy Review* (August 1986), p. 64.

▼ Suggested Reading

FADHIL AL CHALABI, *OPEC and the International Oil Industry: A Changing Structure* (Oxford: Oxford University Press, 1980).

HANS JACOB BULL-BERG, *American International Oil Policy* (London: Frances Pinter, 1987).

EARL COOK, *Man, Energy, Society* (San Francisco: W. H. Freeman, 1976).

PETER COWHEY, *The Problems of Plenty: Energy Policy and International Politics* (Berkeley, Cal.: University of California Press, 1985).

JOEL DARMSTADTER, *Energy Today and Tomorrow* (Englewood Cliffs, N.J.: Prentice-Hall, 1983).

CHARLES EBINGER, *The Critical Link: Energy and National Security in the 1980s* (Cambridge, Mass.: Ballinger, 1982).

FEREIDUN FESHARAKI AND DAVID ISAAK, *OPEC, The Gulf and the World Petroleum Market* (Boulder, Colo.: Westview Press, 1983).

PRADIP GHOSH, ed., *Energy Policy and Third World Development* (Westport, Conn.: Greenwood Press, 1984).

NEIL JACOBY, *Multinational Oil* (New York: Macmillan, 1974).

BRUCE JENTLESON, *Pipeline Politics: The Complex Political Economy of East–West Energy Trade* (Ithaca, N.Y.: Cornell University Press, 1986).

PAUL KEMEZIS AND ERNEST WILSON III, *The Decade of Energy Policy: Policy Analysis in Oil-Importing Countries* (New York: Praeger, 1984).

ROBERT LIEBER, *The Oil Decade: Conflict and Cooperation in the West* (New York: Praeger, 1983).

DON KASH AND ROBERT RYCROFT, *U.S. Energy Policy: Crisis and Complacency* (Norman, Ok.: University of Oklahoma Press, 1984).

WILLIAM LEFFLER, *Petroleum Refining for the Non-Technical Person* (Tulsa, Ok.: PennWell Publishing, 1979).

AMORY LOVINS AND L. HUNTER LOVINS, *Brittle Power: Energy Strategies for National Security* (Andover, Mass.: Brick House Publishing, 1982).

MERRIE KLAPP, *The Sovereign Entrepreneur: Oil Policies in Advanced and Less Developed Capitalist Countries* (Ithaca, N.Y.: Cornell University Press, 1987).

ZUHAYR MIKDASHI, *Transnational Oil: Issues, Policies, and Perspectives* (New York: St. Martin's Press, 1986).

OFFICE OF TECHNOLOGY ASSESSMENT, *Nuclear Power in an Age of Uncertainty* (Washington, D.C.: U.S. Congress, 1984).

JOAN PEARCE, *The Third Oil Shock: The Effects of Lower Oil Prices* (London: Routledge and Kegan Paul, 1983).

ANTHONY SAMPSON, *The Seven Sisters* (New York: Viking, 1976).

IAN SEYMOUR, *OPEC: The Instrument of Change* (New York: St. Martin's Press, 1981).

MARY ANN TETRAULT, *Revolution in the World Petroleum Market* (Westport, Conn.: Quorum Books, 1985).

DANIEL YERGIN AND MARTIN HILLENBRAND, *Global Insecurity: A Strategy for Energy and Economic Renewal* (Boston: Houghton Mifflin, 1982).

4

The Political Economy of Feast and Famine

The economic welfare of people in industrial countries requires access to adequate and reasonably priced supplies of fossil fuels. But for all of the world's people, both rich and poor, energy from food is more vital. Modern agriculture now links together fossil fuel and food energy problems in very complex ways. Without adequate supplies of food, the human machine can't function. And without reasonably priced fossil fuels, many of the recent gains in world agricultural production would be nullified.

Throughout most of human history food production took place almost entirely within the constraints of current solar income. Solar energy was captured by plants through the process of photosynthesis and made available in this form to human beings and animals. Prior to the 1960s, most increases in world food production thus took place because larger quantities of land were already cultivated. But recent increases have resulted from greater productivity of land already under cultivation. Contemporary agriculture has undergone its own industrial revolution, and agriculture is now a high-technology business, with fossil fuel–based fertilizers, pesticides, herbicides, and machinery making each acre under cultivation much more productive. It is estimated, for example, that if the current fertilizer subsidy were removed from contemporary agriculture, world food production would drop by one third.[1]

Aside from the increasingly important fossil fuel–food production link, there are other relationships and similarities between world energy and food problems. Just as global energy interdependence has increased over the last twenty years, a worldwide market for food has also slowly developed. By 1974–75, total world grain trade had grown to 135 million metric tons, which represented about 11 percent of world production. Thirteen years later, 196 million metric tons of grain were traded, representing more than 12 percent of world production.[2] Although this represents a substantial increase in the size of the world market, a significant portion of the world's food, particularly rice, is still consumed in the country of production. But the trend over the last two decades has paralleled developments in the energy market: more food is crossing borders in order to meet needs in the Soviet Union, many of the less developed countries, and several of the OPEC countries.

The world food market has been subject to boom and bust cycles very similar to those that have plagued the oil industry. In the 1960s there was a superabundance of food, and American farmers were paid considerable sums to keep land out of production. But in the late 1960s and early 1970s, a spate of books predicted global starvation because of projected food shortages.[3] Indeed, a world "food crisis" coincided with the first round of oil price increases, and world food stocks dropped to record lows by the mid 1970s. The oil glut of the mid 1980s was accompanied by another period of world food surplus as carry-over stocks of grain reached record levels, much to the chagrin of American farmers. Although a significant reason for fluctuations in world food supplies is weather, the rise and fall of oil prices and the prices of related high-tech agricultural products, such as pesticides, herbicides, fertilizers, and gasoline, also have been clearly related to food prices and availability.

However, there are some very important differences between world energy and food markets. The most important difference is that, quite unlike the quasi-monopolistic world oil market, which has been controlled for most of its history by one group or another, the international food market has been a relatively free one. Demand and supply and related market prices have played a much more important role in world food trade. This is not to say that the market is perfect. For example, the United States controls a little less than 50 percent of world grain exports. Moreover, the governments of industrial countries are heavily involved in subsidizing agricultural production and related exports. But behind the domestic political considerations that may dictate such policies, the ultimate reality is that exporters and importers freely interact in a world food market.

The food market is also different in that it represents a sort of mirror image of the energy market, the exporters of food being the importers of oil and vice versa. The United States is a large energy importer but by far the world's largest food exporter. On the other side of the coin, the Soviet Union, a considerable exporter of energy, is the world's major food importer. Similarly, many of the large Western European energy importers have recently become food exporters, and several significant OPEC countries, including Saudi Arabia and Nigeria, have become big food importers. It is somewhat ironic that many of the industrial countries have been using the products of their complex energy-intensive agricultural technology to gain export revenue with which to pay for the imported oil that has become an essential element of modern farming. It is also ironic that recent gains in world food production have taken place in the highly industrialized countries, where population growth has leveled off. In many of those countries experiencing the most rapid population growth, agricultural production has fallen far short of keeping up with domestic needs, largely because of the difficulties involved in transferring high-technology agriculture to impoverished Third World settings.

Just as it is very difficult to predict precisely the future course of the complex world energy market and when a third crisis might be triggered, it is also difficult to make precise predictions about the world food market. The successes or failures of the moment tend to color long-term perspectives. The future shape of the market, however, will be determined by three sets of factors: objective need for food, production possibilities, and patterns of effective demand.

The objective need for food is a function of future population growth and related dietary requirements. Little increase in demand can be expected in the industrialized countries, where population growth has subsided. Rather, the greatest future demands for food can be expected in the less developed areas of the world, where populations are growing very rapidly. The availability of food on a world market permits less developed countries to sustain populations much larger than their own farmers can feed. In fact, in many of these countries a massive migration of inhabitants from rural to urban areas holds down food production precisely when rapid population growth is creating new demands. In addition, modernization creates more pressures on food production, as people shift from grain consumption to eating more meat, which is a much less efficient use of agricultural resources.

At the present time, world production capabilities would be adequate to meet minimum nutritional needs for all human beings in an average year if food were distributed in a more egalitarian manner. But there are limits to future production possibilities. Lester Brown has called attention to some of these more pessimistic aspects of the world food problem. Although stockpiles of grain were at record levels in the mid 1980s, because of increases in population those stocks represented about the same number of days of world consumption as they did during the food crisis of the mid 1970s. In addition, over the long term, soil erosion and lack of water for irrigation threaten to limit future production as more marginal land is brought under the plow.[4] But most important of all, the tight link to energy supplies made necessary by heavy reliance on fertilizers, pesticides, herbicides, and mechanized equipment ties food prices and availability very closely to the world energy situation.[5] And climatologists point to evidence that the last few decades have been extremely propitious for agricultural productivity in the temperate regions of the world and that changes in weather and climate could easily create a very different food situation in a very short period of time.[6]

Given large periodic food surpluses in the industrial countries, how is it that starvation and malnutrition have been so obvious in parts of Latin America and Asia and in substantial portions of Africa? The answer is found in changing patterns of effective demand for food. Effective demand results from objective need for food combined with cash to purchase it. The sad fact is that in those countries in which objective need is rapidly growing people don't have the purchasing power to buy the expensive products of high-technology agriculture. Thus, American farmers have been paid to keep land out of production while millions have perished in Africa and elsewhere, largely because the starving don't have the cash to influence the world market. In the absence of a dramatic change in the international distribution of purchasing power or in food aid by the industrial countries, in the future in dozens of countries starvation likely will be common because of population growth, urban migration, and a highly inegalitarian distribution of wealth.

▼ Basic Food Requirements

A number of different nutrients found in a variety of foods are essential for human well-being. Normally, human food needs are defined in terms of the number of calories and amount of protein consumed on a daily basis. But the matter is really more complex. People also require various kinds of vitamins and minerals to maintain their health.

The human machine gets energy by burning carbohydrates and fats. The amount of potential energy in food is measured in kilocalories. A *kilocalorie* is the amount of energy that must be expended to raise the temperature of one kilogram of water by one degree centigrade. Diet guides and books dealing with dietary requirements substitute the word *calorie* for *kilocalorie*. This convention is also adopted here. Human beings vary tremendously in the number of calories required according to their age, sex, and the climate they live in, but the minimum requirement is about 1800 per day for older females, and the maximum is about 3000 for young males. Regionally, the average North American requires about 2650 calories, West Europeans require about 2550 calories, Africans about 2350, and Latin Americans about 2400.[7] In general, people in the industrially developed countries consume significantly more calories than they require, whereas those in less developed countries make do with considerable shortfalls.

Proteins are also essential to human health because much of the body is made up of them. There are millions of different kinds of proteins, but most of them are derived from about twenty different amino acids. Protein is essential for growth and physical development and maintains and repairs tissue. Various types of disease and retardation result from protein deficiency. Foods vary in the nutritional value of the protein they contain, and it is therefore difficult to be exact about daily requirements. If eggs, for example, are indexed at 100 in terms of net protein usability for human beings, fish would weigh in at 83, meat 82, milk 75, vegetables 70, grains 56, and peanuts 48, for similar quantities consumed. Thus, in very general terms a person would have to eat twice as many peanuts as eggs to net a similar amount of useful protein. Not all the proteins required are found in all foods, however, and a varied diet is required for optimal health. When all factors are considered, protein requirements vary from about 46 to 54 grams per day.[8] As is the case with calories, consumption of protein around the world is far from equal. People in the industrial countries consume about twice the minimum protein required and people in less developed countries often get less than the minimum and in less desirable forms. In the industrial market economies, for example, roughly 32 percent of protein comes from animal products, whereas in Africa only 7 percent comes from those sources.[9]

Although not as commonly mentioned in international reports and statistics, vitamins and minerals are also important components of the human diet. About thirteen vitamins are considered to be essential. Only small amounts of them are needed, but should significant vitamins be lacking, various diseases or death could result. There are also seventeen essential minerals that human beings extract from the food that they eat, and if they don't get adequate supplies similar diseases and problems result.[10] Normal diets in the industrial countries contain the required vitamins and minerals, and if people have deficiencies they take pills to compensate for them. But in less developed countries, simple and monotonous diets based on grains or rice often lack the required variety of vitamins and minerals.

There are many different ways of categorizing the foods that produce the carbohydrates, fats, proteins, vitamins, and minerals needed in the human diet. Lester Brown has divided them into six categories: cereals; roots and tubers; fruits, nuts, and vegetables; sugar; fats and oils; livestock and fish.[11] The three species of cereals—wheat, rice, and corn—are by far the most important food plants, accounting for nearly half of the food energy consumed around the world. Wheat is the dietary staple in the cooler regions of the world, whereas in the more tropical areas rice is the mainstay of the diet. Rice is considered the world's most important food since it is essential to the diets of about one half of the world's population. Corn is the other important cereal, but large quantities of it are fed to animals, and the nutrients in it often make their way indirectly into the human body through meat. Direct consumption of corn is limited mainly to Latin America and parts of Africa, although some forms of corn are directly consumed in the United States.

The other half of world food energy comes from a wide variety of substances, none of which approach the cereals in importance. Meat and fish, foods that are very common in industrial countries, account for only about 11 percent of the food consumed worldwide. These are luxury items that the bulk of humanity living in the less developed countries cannot afford. The root and tuber family produces a large quantity of potatoes, but their water content is very high (75 percent) and their bulk somewhat restricts their movement in international trade. Potatoes also have a relatively low protein content and thus they are not desirable as foods in protein-deficient countries. Other types of roots and tubers, including yams, cassava, and sweet potatoes, are important components of the diets of many people in less developed countries.

Fats and oils are provided by legumes, which also are a source of protein. Soybeans and peanuts are grown primarily as a source of oil for industrial processes but also provide the base for margarine and related food products. Processed soybeans are becoming more important in the human diet as a protein supplement. In the early 1970s the world soybean market was dominated by the United States and China, but more recently Brazil has become a very significant producer and exports considerable quantities of soybeans to Japan.

The remaining two categories encompass crops such as fruits, vegetables, and also sugar. Fruit and vegetable production is well dispersed around the world, and these commodities do not figure significantly in international trade. Sugar cane is grown primarily in less developed countries in the more tropical areas of the world and is an important export crop. But world sugar prices were depressed in the 1980s because of a growing supply and slackening demand. Various high-technology substitutes for sugar have been developed in industrial countries, and the use of these "diet" substances has dampened sugar demand. In addition, the United States carefully controls the amount of sugar that can be imported in order to protect producers of high-priced domestic sugar beets.

▼ Production and Markets

Prior to the 1970s, domestic agricultural policies were not considered relevant to international relations. Although agricultural commodities have always crossed national borders in international commerce, prior to the 1970s the quantities traded were small and limited to discretionary items and gastronomic delicacies. It has only been in the last two decades that a global food market and related food policies have developed.

Before a world market for agricultural commodities existed, countries were at the mercy of nature in meeting domestic nutritional needs. Populations were trimmed by starvation and malnutrition if harvests fell short of domestic needs. There was no world food market, international granary, or famine relief agency to meet the needs of the starving. Famines were frequent but often localized affairs. Many of them passed unnoticed by Western historians because they occurred in remote areas of the world. But there is plenty of evidence that the Malthusian grim reaper was regularly at work. In 1878, for example, Cornelius Walford published a study chronicling more than two hundred famines in Great Britain between the time of Christ and 1850. He also listed more than 150 major famines that had occurred in other parts of the world.[12] Another historical study recorded nearly one famine per year in some part of China for the two-thousand-year period preceding 1911.[13]

The growth of a world food market has been both a positive and negative development. On the positive side it has freed many countries from the necessity for complete food self-sufficiency, thus making labor forces available for industrial development. It has also permitted food exporters such as the United States to utilize more fully their agricultural resources. The world market also can smooth out the impacts of localized sporadic bouts of bad weather. Countries that are adversely affected can turn to the world market for short-term relief. But on the other hand, the existence of a world market has seduced leaders of many less developed countries away from agricultural self-sufficiency and exposed them to potentially serious problems should reasonably priced food not be available in the future. As populations have swarmed from rural areas to the cities, these countries have become permanent customers in the food market—permanent, that is, until the hard currency with which to buy these commodities disappears. Finally, while a reduction of local famines undoubtedly has been associated with greater availability of food for sale, as the world market has grown the prospect of global famine brought on by simultaneous crop failures in two or more exporting countries has become a much more distinct possibility.

The contemporary world food market now receives about 15 percent of world production of wheat and course grains but only about 4 percent of the world's rice crop. Since about 50 percent of the world's calorie consumption and a slightly smaller percentage of protein consumption come from cereals, the bulk of trade in food takes place in these categories.

Wheat is a dietary staple for large numbers of people living in temperate climates. Wheat grows well where winters are cool and summers are hot. It does not thrive in the tropics, where various plant diseases attack it. Although the Soviet Union and China have both been large importers of wheat over the last decade, they also are currently competing for honors as the world's largest producer, the United States coming in a distant third. Wheat is the most important commodity in international agricultural trade, with about one fifth of total production regularly crossing national borders. Trade is much more highly concentrated than is production, with the United States regularly accounting for about 40 percent of all exports. Canada, the European Economic Community, Oceania, and Argentina account for most of the remainder of wheat exports.

Rice is essential to the diets of the half of the world's population that lives in the more tropical areas. Although the United States is not the world's largest producer of rice, it is a major exporter. Most of the big rice producers, such as China and India, consume domestically most of the rice that is harvested. Between the two of them, China and India produce more than half the world's rice. The 3 or 4 percent of the world's rice crop that enters the international market is exported by Thailand, the United States, Pakistan, and China.

The rest of the cereals, including corn, fall into a category called coarse grains. Corn is the third largest cereal crop grown and the only major one that is indigenous to the western hemisphere. Corn grows well in the American midwest, and about half the world's production comes from there. About 15 percent of all coarse grain produced is exported. The United States produces about one quarter of the total supply of all coarse grains, followed by the Soviet Union and China. Trade in coarse grains is highly concentrated, with the United States accounting for about 70 percent of the total. Argentina is the only other exporter to have a significant share of the market. (See Figure 4-1.)

The contemporary world food market is thus dominated by exports from the United States. Whereas the production of cereals is spread among many countries, exports are not. Both the wheat and coarse grain markets are dominated by the United States, and so little rice is exported as to be insignificant. When everything is added together, the United States produces about 30 percent of all cereals and comes very close to accounting for one half of all sales in the world market. Exports of agricultural commodities account for about 20 percent of U.S. export earnings and can be looked upon as both a strength and weakness of the American economy. These exports are essential to preserving some semblance of balance in U.S. trade figures. But farmers have become dependent upon foreign markets in order to sell their products, and keener foreign competition combined with lagging demand may substantially cut the U.S. share of the world market in the near future.

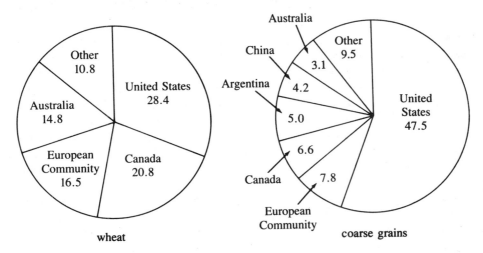

Figure 4-1 Major wheat exporters and major coarse grain exporters in million metric tons (1986–87 estimates).

SOURCE: U.S. Department of Agriculture World Grain Situation and Outlook, December, 1987.

▼ Constraints on Food Production

Four factors will determine the long-term ability of farmers to feed the world's growing population. The most obvious is availability of land and water, since only limited supplies of both can economically be made usable. The world's farmers already have staked claims to the most productive land available, and that which remains to be exploited is marginal—less productive and risky to bring into production. Irrigation is commonly used to bring marginal land up to standards, but fresh water is also scarce, and desalinization of ocean water is very expensive. And in many industrializing countries, there is heavy competition between agriculture and industry for the water required by both sectors.[14]

Modern agriculture is very energy-intensive, and the future price of fossil fuels represents another factor that may constrain production. Over time, less food production will come from solar energy and relatively more will come from increased reliance on mechanized farming. Industrialized agriculture uses modern equipment, fertilizers, pesticides, herbicides, and frequently irrigation, all of which require substantial quantities of petroleum, electricity, and natural gas. When future energy prices begin to rise, the cost of farming and the food produced will likely rise in step.

Technological developments in agriculture must also be weighed when assessing the future of world food production. When Reverend Thomas Robert Malthus wrote his famous *Essay on the Principle of Population as It Affects the Future Improvement of Society,* he was ignorant of future technological innovations that would have a great impact on agricultural productivity.[15] He projected a nasty and brutish future for the human race, as exponentially growing human populations overwhelmed the amount of land available. But technological innovations in farming have been responsible for increasing agricultural productivity and thus proving the Reverend Malthus at least partially wrong. Scientific research has resulted in higher-yield varieties of crops that, when properly fertilized and irrigated, make each acre of land more productive. New developments in genetic engineering hold promise that major breakthroughs in agricultural productivity could take place during the early years of the twenty-first century.[16]

The last, and most important, factor determining future food production possibilities is climate. Research indicates that history is filled with regional and global changes in temperature and rainfall. Many climatologists believe the last few decades of agricultural production in Europe and the United States have resulted from some of the best growing conditions that have ever existed in those regions. The implication of their findings is that more historically normal conditions can be expected to return to these regions, thus limiting future harvests.[17] Other researchers have noted a buildup of carbon dioxide in the earth's atmosphere and the potential for a related greenhouse-type warming, which could also be expected to upset traditional temperature and rainfall patterns. Although much of this research is still fragmentary, climatological change could seriously disrupt world food production if it occurs, and in view of recent heat and drought in the United States certainly is a factor to be closely watched in making projections about future food availability.

The supply of arable land has been considered the primary factor potentially limiting growth in food production ever since the Reverend Malthus called attention to the seemingly limited supply in 1798. Over the years, equipment for assessing the limits of arable land has become much more sophisticated. Orbiting satellites now regularly photograph cropland, and there is much less mystery about the quantity and quality of unfarmed but potentially usable land.

Assessments of the world's supply of land seem to have come to rather optimistic conclusions. Studies estimate that there are about eight billion potentially arable acres, fewer than half of which are now being farmed.[18] The availability of this land varies by continent. It is estimated that only one quarter of the potentially arable land in Latin America is now being farmed, slightly more than one third is being farmed in Africa, one half is being farmed in the United States, two thirds is being farmed in Asia, and three quarters is being farmed in the Soviet Union. Most of the arable land in Eastern and Western Europe is already considered to be under the plow. In Australia and New Zealand vast tracts are yet to be brought under cultivation.[19]

These estimates must be viewed with some suspicion; usable land may not be as abundant as it seems. Most studies overestimate the utility of unfarmed land by underestimating the difficulties that would be involved in bringing it under cultivation. Much of the land that is not now being farmed outside of the United States is land on which farming in not likely to be economically viable for the foreseeable future. The experience of generations of farmers has determined which parcels are economically viable and, even in the face of intense population pressures, not much of this potentially usable cropland can be counted on for farming. More than half of it lies in tropical regions. Brazil, for example, is attempting to push back its tropical rain forest in order to create new farming opportunities for its burgeoning population, but farmers have been bitterly disappointed as much of the rain forest lies atop very poor soil that will not support normal agriculture.

There are many other reasons that much of this potentially arable land will never be turned under the plow. Most marginal land is found in areas where rainfall and temperature are very erratic. Two or three years of good harvests may be followed by protracted droughts, bankrupting any farmers who might settle there. Much of this land is best suited to grazing, having a much lower potential for crop production than land now being farmed. In vast stretches of Africa, the tsetse fly would devour anyone who tried to settle on what is officially defined as potentially arable land, and periodic locust plagues damage much of the land already under cultivation. Finally, a great deal of this marginal land is geographically remote from markets and population centers. Developing a transportation infrastructure to bring crops to market would be prohibitively expensive for many of the less developed countries where this land is found.

Fortunately, this marginal land probably will not be the most immediate factor limiting future food production, because recent increases in production have resulted from obtaining better yields on existing cropland rather than from bringing new land under cultivation. Until the 1950s, the amount of land under cultivation grew by almost 1 percent per year, but since that time additions to cropland have been relatively small. It is estimated that additions to arable land will total only about 4 percent for the entire period from 1975 to 2000.[20] About one third of the world's population lives in countries where the amount of land devoted to crops is actually shrinking. Ireland has lost nearly one third of its cropland since 1960, Japan about one fifth of its cropland, and France about one seventh.[21] In these countries land is often lost to urbanization, highways, and industrial activities, but increases in agricultural productivity on existing land have compensated.

World food production is expected to continue increasing largely because of application of more fertilizer. In 1950 only 14 million metric tons of chemical fertilizer were applied to agricultural land. By 1984 chemical fertilizer consumption had skyrocketed to 121 million metric tons; consumption is projected to reach 220 million metric tons by the year 2000.[22] With the aid of this energy-intensive subsidy, yields per acre are expected to rise by nearly 25 percent over the next fifteen years. Thus, world food production is projected to grow dramatically, with a 90 percent increase between 1970 and 2000.[23] This increase in production will be accomplished with a per capita decline in arable land in all regions of the world.[24]

Water may ultimately play a more important role in limiting world food production. Rainfall is scarce in many of the untilled regions of the world, and the potential for additional irrigation is limited. All of the farmland in Egypt is now under irrigation. In China, Pakistan, Taiwan, Korea, the Netherlands, Israel, and Japan more than 40 percent of all cropland is now being irrigated.[25] But irrigation is expensive, and unpolluted water is a finite resource that is increasingly in scarce supply.

The crux of the water problem is that dependable streamflow in many parts of the world is not great enough to meet projected needs. In the United States, for example, more than one third of all streamflow is being used. It is estimated that by the year 2000 all dependable U.S. streamflow will be used at least once on its way to the oceans. Although water that is withdrawn is not usually completely consumed, it generally is returned to streams laden with pollutants, thus creating problems for those wishing to use the water further downstream. About 90 percent of the water used in commercial and residential applications is eventually returned to the streams from which it is withdrawn. But only 40 percent of water used for irrigation makes its way back to the streams.[26]

Much of the world's irrigated land is supplied by wells, which in many cases are akin to nonrenewable resources. Continuous pumping lowers the water tables much more rapidly than nature can replenish the stock of water. Tube wells used for irrigation in India, for example, have a very short useful life since these underground resources dry up quickly. Many new wells have to be drilled just to replace the old ones that are drying up.[27] Disputes over irrigation water are also a source of conflict among nations. Use of Colorado River water, for example, has been a major source of controversy between Mexico and the United States. As farmers in the United States have withdrawn larger quantities for irrigation, Mexico has been left with only saline water on its side of the border.[28]

Development of new technologies and the possibility of significant changes in climate that will impact future food production are long-range factors that are much more difficult to project. Since increases in food production are now closely tied to energy-intensive technologies, future food production will be much more dependent on the price of gasoline, electricity, fertilizer, herbicides, and pesticides than it will be on the availability of additional land. In the early 1970s, one hectare of arable land supported about 2.6 persons. By the year 2000 each hectare will be supporting 4 persons.[29] This productivity will be sustained by applying an average of 145 kilograms of fertilizer to each arable hectare of land, as opposed to only 55 kilograms per arable hectare in 1971–75.[30]

Large-scale introduction of mechanized farming has done for agriculture what the large-scale utilization of fossil fuels in mechanized production did for industry. Machinery has replaced human labor. As a result, in the United States only 2 percent of the population now produces enough food to feed the rest, as well as considerable numbers of people in other countries.

The penetration of the industrial revolution into agriculture has been very uneven around the world, and at present the industrial countries reap the greatest benefits from the new technologies while the less developed world trails far behind. For example, in Europe 223 kilograms of fertilizer are used per hectare of arable land, and 94 kilograms are used per hectare in the United States; in Africa, by contrast, only 19 kilograms of fertilizer are used per arable hectare. In the densely populated Netherlands, food self-sufficiency is maintained by applying more than 750 kilograms of fertilizer per hectare. In several African countries less than 1 kilogram of fertilizer is applied per arable acre.[31] Thus, the industrially developed countries have parlayed their competitive edge in fossil fuel–based technologies into agriculture, and, even though land is scarce, many have become significant exporters of agricultural commodities.

The first two energy crises were accompanied by parallel dramas on the world's farms, as the price of energy-intensive inputs into agriculture rose apace. In fact, in the twenty-three years preceding the oil crisis of 1973–74, fertilizer use was growing at about 7.5 percent per year. After the second energy crisis in 1979–80, fertilizer utilization grew at only 2.5 percent per year.[32] Data on the growing energy requirements for global agriculture are difficult to obtain because of the impossibility of collecting such figures from less developed countries. But the changing energy requirements of American agriculture give some idea of the impact of new energy requirements associated with the mechanization of world agriculture.

Several studies of the fossil fuel energy component of American agriculture were carried out in the 1970s, and the trends noted in them have persisted well into the 1980s. One study tracked the increase in energy inputs into American corn production over a twenty-five-year period (1945–70). It found that the typical acre of corn required an input of 926,000 calories of fossil fuel energy in 1945. By 1970 the input had jumped to 2,897,000 calories per acre. In the earlier year, 3.7 calories of food energy were returned for every fossil fuel calorie invested, whereas in the latter year the yield ratio had dropped to 2.8 food calories for each calorie of fossil fuel used. Worldwide, 276 million barrels of oil equivalent were used in world agriculture in 1950. By 1985 this figure had grown to 1.9 billion barrels. It is estimated that .44 barrels were used to produce one ton of grain in 1950, as opposed to 1.14 barrels in 1985.[33]

This mechanization of agriculture presents potential future problems because industrialized agriculture is energy intensive, and thus its products are very expensive. It now costs the U.S. economy about ten calories of fossil fuel to put one calorie of food energy on the table.[34] Not only are large quantities of energy required on the farm, but additional energy is used in food processing, packaging, freezing, transportation, and so on. In addition, industrialized agriculture stresses meat and poultry production, which represents a waste of both protein and calories. Feedlot beef in the United States, for example, is only 6 percent efficient in transforming the protein in feed to protein in edible meat. It should be obvious that a world of five billion people cannot be fed by an industrial agricultural system such as that found in the United States, because there simply is not enough money, land, energy, or water in most developing countries to make such a system work there.[35]

These figures have very serious implications for the world's malnourished people in less developed countries. World food prices move in tandem with energy prices. After a substantial pause in the 1980s, energy prices are likely to begin to rise again, and food prices should also rise considerably. The average per capita expenditure for food in the United States is now close to $1500, a figure well beyond the average per capita income in most less developed countries. While industrial agriculture can continue to work for the rich countries, it is unlikely to be successfully transplanted to the world's poorer nations.

Mechanized agriculture presents additional long-term problems. Heavy fertilizer use and irrigation are both associated with the problem of diminishing returns. Over time, irrigated and fertilized soils tend to become saline and have to be rested frequently if they are to continue to be productive. There is evidence that corn yields are maximized in the United States when about two hundred pounds of nitrogen fertilizer are applied per acre. Worldwide in 1950, each ton of fertilizer applied yielded about 46 tons of grain. By the early 1980s, this figure had dropped to about thirteen tons of grain per ton of fertilizer.[36] In many places there is now very little additional yield per acre when more fertilizer is added. In the less developed countries, however, diminishing returns are not yet a problem. For them, higher fertilizer prices in the face of huge foreign debts has led to restrictions on imports and less agricultural production.

In the more distant future, post-industrial technologies applied to agriculture will certainly be able to augment the world's food supply. Advances made in biotechnology in the 1980s may be applied to agricultural production as early as the late 1990s.[37] Genetic engineers may develop new strains of crops capable of greater yields and improved farm animals that dwarf present varieties. But it is uncertain whether such developments will change the world food situation in a fundamental way. The new technologies will undoubtedly be responsive to agricultural interests in the industrialized countries, where overproduction is already a problem, and will be only slowly developed for those areas of the world in greatest need of more food.

▼ Changing Climates

The factors just discussed that shape future food production are understood and, to a certain extent, can be controlled by human efforts. But there is presently little that human beings can do about the weather. It is well established that climate, which represents an ultimate constraint on agricultural productivity, has changed significantly over the centuries. Reid Bryson, for example, points out that since 1700, all thirty-year periods were colder than the one between 1930 and 1960. In addition, 90 percent of the last million years has been significantly colder than the present era.[38]

Weather changes slightly from day to day and year to year. Short-term fluctuations in temperature and rainfall regularly take place throughout the world. Changes in climate, however, are more fundamental long-term propositions that can make some of the most fertile agricultural regions into barren wastelands. There is no doubt that major climate changes have taken place in the past, and there is little reason to think that they won't take place in the future. And the dynamics of climatic change are not yet very predictable. It is clear that many of the prime agricultural production areas of the world were covered by glaciers at various times in the past; it is uncertain when, if ever, such a situation might recur.

Tracking the historical climate record is very much like a detective game. Not only is it difficult to estimate past temperature changes with any precision, but tracing the impact of small changes on wind patterns and rainfall is even more difficult. Bits and pieces of information about past climates come from a wide variety of sources. Some evidence of past conditions can be extracted from glaciers and polar ice caps. Many of these ice layers are thousands of feet thick and represent a history book for scientists as they core down to great depths. Evidence of what was once airborne atmospheric dust, various chemicals, volcanic activity, and so on has been deposited in the glaciers. In addition, the extent of past glacial coverage can be gauged through geological evidence. Other information can be recovered from plant and animal fossils. As the earth alternately warmed and cooled, tree lines and various forms of wildlife moved in response to changing temperatures. By examining fossil records at various depths, climatological histories for different regions of the world can be constructed. For more recent periods, the width of tree rings offers evidence of both temperature and rainfall. Even the ocean floor contains a treasury of information about past climatological conditions. The soil on the ocean bottom is a repository of fossil life from eons in the past. Again, depending on the depths of core samples taken, the forms of fossils found are indicative of ocean temperatures during various historical periods.

Three things could happen to change climate and thus upset established patterns of global food production. There could be a change in the amount of energy reaching the earth's atmosphere from the sun; the ability of the earth's atmosphere to transmit this energy could change; and there could be a change in the earth's *albedo,* or the reflectivity of its surface.[39] The so-called solar constant, the amount of energy given off by the sun, is not really constant. Very small changes in the amount of solar productivity could cumulate and have a major impact on the earth's climate. In addition, the earth is not always the same distance from the sun; this could also affect the amount of solar energy reaching the atmosphere.

Climate shifts are much more likely to be triggered by the expected changes in the transmittance of the earth's atmosphere. About 30 percent of incident solar energy is immediately reflected back into outer space. The rest of it is absorbed by the atmosphere or by the earth. Changes in the amount of particulate matter in the atmosphere or in its chemical composition can alter normal patterns of absorption and thus the average temperature of the earth. Even small temperature changes of one or two degrees centigrade can change rainfall and snowfall patterns and contribute to changes in the albedo. In recent years the winter snow cover in the northern hemisphere has been extremely large by historical standards, leading some to believe that major changes are now under way.

Past climatological changes have been triggered by various natural phenomena. There is good evidence that major volcanic eruptions have triggered short-term, if not long-term, changes in climate and weather. Volcanic explosions on the scale of the Tambora eruption of 1815 or the Krakatoa explosion of 1883 can spew so much material into the atmosphere as to change dramatically the transmittance factor. The Tambora explosion is estimated to have lowered global temperature significantly for one year and adversely affected harvests around the world for two years—the approximate length of time that it took for the particulate matter to settle to the earth.[40] It is speculated that global volcanic eruptions are cyclical, that they are correlated among the various regions of the world, and that these eruptions might well be responsible for triggering changes leading to ice ages. Given projected population pressures against limited food reserves, future substantial volcanic eruptions could directly result in tens of millions of deaths from starvation.

When much fragmentary evidence is added together, two trends can be discerned that seem to be pulling in opposite directions. On the one hand, geological records yield strong evidence of an imminent global cooling. Reid Bryson concludes his extensive study of the historical record by relating his concern that the world is moving toward cooler temperatures, for which the human population is ill prepared. He argues that the historical record indicates that a global cooling is likely and that such climate changes can take place very rapidly. Temperature changes themselves may not be the main problem, however. Rather, the accompanying changes in wind patterns and rainfall could wreak havoc on a more densely populated planet.[41]

On the other hand, there is concern that mankind's industrial activities, which create large quantities of carbon dioxide and other gases, are changing the composition of the earth's atmosphere and thus the transmittance factor. A greenhouse effect, described in greater detail in Chapter 5, is resulting from a continuing buildup of carbon dioxide in the earth's atmosphere. It is thought that there were about 290 parts per million of carbon dioxide in the atmosphere in the year 1800 and that there will be 590 parts per million by the year 2060. Between 1958 and 1980, the carbon dioxide content of the atmosphere, measured at Mauna Loa, Hawaii, increased from 316 to 336 parts per million. Studies indicate that such changes could lead to a warming of more than two degrees centigrade by the year 2040.[42]

In summary, many imperfectly understood forces are operating to change world climates. Geological evidence indicates the imminence of another ice age; chemical evidence of carbon dioxide buildup indicates a significant global warming trend. No one can predict which set of forces will dominate over the next few decades, but there is general agreement that this combination of factors could produce temperature and rainfall extremes—that is, natural disasters—much greater than those experienced over the last fifty years.[43] If these changes do take place, it is likely that crops planted in the more marginal areas of the world will suffer the greatest damage. Thus, the northern latitudes of the Soviet Union and Canada might well be at risk. But the greatest risk will be borne by those less developed countries that continue to be major food importers, since any weather-related declines in world food production would likely lead to much higher food prices and nationalistic food policies restricting food exports from the major producing countries.

▼ U.S. Interests and World Markets

Recently the world food market has had a surplus of grain. Yet it is estimated that more than a billion people have been chronically undernourished, and hundreds of thousands have recently died of starvation in several African countries.[44] Clearly, there is something perverse about a world food market that permits large numbers of people to be malnourished or to starve while farmers are encouraged to hold land out of production in the exporting countries.

Just as is the case with energy, perspectives on the food problem have varied in cycles with the exigencies of the moment. In the 1960s, for example, solving the "farm problem" in the United States involved devising ways to deal with a massive glut of grain. There was clearly too much production for the domestic market, and government-sponsored programs were established to move food abroad at concessional prices. Largely because of the political strength of the farm lobby, agricultural prices were supported at levels well above what an unregulated market would dictate.

Perspectives on the world food situation changed dramatically in the early 1970s, when the problem of glut was suddenly transformed into a world crisis of scarcity. Because of bad harvests, rising energy prices, and additional food demand, world carry-over stocks of grain plummeted from 183 million metric tons in 1971 to a low of 140 million metric tons in 1974, enough of a stock to support only about forty days of world consumption. The latter years of the decade were spent rebuilding depleted stocks, and American farm exports to new customers rose considerably. During the 1980s, however, the food situation came full circle. A glut of production in the major exporting countries and stagnant demand in many of the importing countries led to stockpiles of 340 million metric tons of grain by 1987.[45]

Besides the effects of a four-year run of mediocre harvests, the entry of new customers tightened the world food market in the mid 1970s. The Soviet Union was the most important of these new customers, at least in terms of opening up a new market for American farm exports. The Soviet Union has always been one of the world's top producers of wheat and course grains, but the expansion of Soviet production in the 1960s took place in the so-called virgin lands in Siberia. This marginal land produces widely fluctuating harvests, depending on temperature and rainfall. In the good years—such as 1986, when the Russians produced 210 million metric tons—there is little need to purchase grain abroad. But in the bad years—such as 1975, when production was only 140 million metric tons—the Russians are forced to import large quantities of grain from abroad.[46] Partially in response to internal political pressures for more reliable food supplies and more meat in the Soviet diet, Russian leaders took the country into the world food market in a big way in the mid 1970s. The United States captured a large share of that market until President Carter imposed an ill-fated grain embargo in response to the Soviet invasion of Afghanistan. In spite of the embargo, subsequently lifted by President Reagan in response to pressures from American farmers, the Russians remained good customers for American grain well into the 1980s, when better harvests caused them to cut purchases from the United States.

Political decisions in China in the late 1970s temporarily brought that giant customer into the market. The Chinese decided to revamp agriculture very dramatically as part of a more general liberalization of the Chinese economy. Although the agricultural reforms seem to have paid off in the long run, in the short run the dislocations involved turned China into a major customer for American grain. In 1975, the Chinese produced 285 million metric tons of grain. By 1986, production was up to 391 million metric tons, and there has been little further need for U.S. imports.[47]

The OPEC countries, ranging from the wealthy banker countries to their less developed colleagues, all significantly increased food imports as part of a spending spree that accompanied higher oil prices. For example, Saudi Arabian grain imports jumped from 482 thousand metric tons of grain in 1974 to 5.6 million metric tons in 1982. Given readily available imports, Saudi domestic food production per capita dropped precipitously over the OPEC decade. Several other OPEC countries joined the ranks of major food importers, particularly when measurements are made on a per capita basis. (See Table 4-1.)

The food market was also influenced in the 1970s by large purchases from the less developed countries that indirectly benefited from the OPEC oil price windfall. The massive flow of petrodollars to the Middle East resulting from the two rounds of oil price increases created a recycling problem, which was partially resolved when banker countries made large deposits in multinational banks. The banks, in turn, found eager customers for new loans in the financially strapped less developed countries. Although this solution to the recycling problem helped to create a world debt crisis, it also created immediate purchasing power with which certain LDCs increased food imports.

Many less developed countries going through the early stages of industrialization experience a large-scale migration of farmers to urban areas in search of more

Table 4-1 Major food importers. Figures represent thousands of metric tons of cereals.

	1982	1985
USSR	40,108	43,251
Japan	24,336	26,720
China	20,365	10,394
Egypt	8655	8903
Spain	7402	4183
Italy	6506	7052
Belgium	6307	5322
Saudi Arabia	5584	5036
Korea	5538	6826
W. Germany	4977	6482
Netherlands	4843	5252
Poland	4566	2396
Brazil	4492	4857
United Kingdom	3943	3521
Algeria	3831	5270
Portugal	3504	2204
E. Germany	3313	2083
Iran	3183	4479
India	2818	85
Venezuela	2575	2793

SOURCE: U.N. Food and Agricultural Organization, *FAO Trade Yearbook 1986.*

lucrative jobs. In the absence of concerted government efforts to keep people down on the farm, domestic agricultural production can fall considerably. These production problems are often compounded by government regulations and subsidies. Government-controlled prices paid to farmers are frequently kept unrealistically low in the interest of maintaining the political support of urban populations. And on the other hand, generous government subsidies are used to keep the social peace by lowering prices of basic agricultural commodities, thus spurring consumption among the urban masses. In Egypt a generous food subsidy program has cost the government more than one billion dollars annually and ranks with defense and government employees' salaries as one of the three biggest items in the Egyptian budget. These subsidy programs distort agricultural development by artificially depressing domestic farm prices while increasing reliance on imported food. In recent years, Egypt has regularly been one of the world's four largest food importers.

This influx of new customers into the world food market in the 1970s resulted from the coincidence of factors that are unlikely to be repeated in the near future. Many of these conditions, which created significant new food demand in the 1970s, are now operating to stabilize or diminish demand for imports. The drop in oil prices and related financial worries for OPEC countries have made them much more cautious customers. The less developed countries are striving to cope with huge debts and are looking for ways to avoid market purchases. China has recovered from the turmoil of agricultural reorganization and imports very little grain. The Soviet Union has experienced better harvests and has been able to cut back on food imports, at least temporarily. Although a series of poor harvests has created major starvation problems in several African countries, the concessional food relief that has been channeled to those countries has hardly made a dent in world food surpluses.

The coming decade does not therefore appear promising for American farmers. American farmers were aided by U.S. dominance of a growing market between 1964 and 1985. Whereas agricultural exports averaged only 5.4 billion dollars in the period 1950–64, by the mid 1980s such exports brought in between 35 and 40 billion dollars annually.[48] But the international market for American surplus has become stagnant, with many of the new customers of the 1970s curbing their purchases (see Figure 4-2). And no new customers appear to be on the horizon. Furthermore, the Soviet Union and China have absorbed about one quarter of recent U.S. exports, thus setting up a peculiar symbiotic relationship between Communist governments and capitalist farmers.[49] Even Western Europe, which used to be a major market for U.S. exports, is moving beyond self-sufficiency and becoming a major exporter. Saddled with a large domestic surplus, which in previous years has been exported to a growing international market, there seems to be little prospect of immediate relief for American farmers, unless repeated drought conditions resolve the problem by driving many farmers out of business.

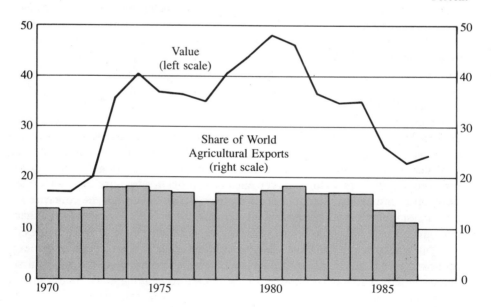

Billions of 1982 dollars

Percent

Figure 4-2 U.S. agricultural exports.

Note—U.S. value for 1987 estimated; world value not available.

SOURCE: *Economic Report of the President 1988.*

The long-term dilemma now facing the United States is that surplus agricultural commodities are produced by market distortions that price American farm products out of the reach of most potential customers, and competition in the food trade is heating up. As a result, the contribution of agriculture to U.S. trade is diminishing. U.S. agricultural exports peaked in 1981 at nearly forty-four billion dollars before dropping back to near thirty billion dollars in 1986.[50] These exports then accounted for more than one fifth of all foreign exchange earnings and farm income.[51] New competitors in the world market, such as Argentina, are able to sell wheat at considerably lower prices than can the United States.[52] During the grain glut, American farmers, saddled with high costs of production, lost as much as seventy cents on each bushel of corn and one dollar on each bushel of soybeans sold.[53] Thus, a veritable feast of grain has been and will be grown in the United States, but at prices too high for the large potential market to afford.

The huge federal deficits of the late 1980s focused attention on the cost of farm price support programs and the related costs of carrying surplus stocks. It is estimated that farm subsidies cost U.S. taxpayers about eighty billion dollars over the 1987–89 period.[54] Some of the resulting surplus can be disposed of through the Food for Peace program (Public Law 480) which, since 1954, has provided food to politically acceptable needy countries through long-term loans and grants. A related program partially subsidizes the cost of agricultural exports. Taken together, these programs cost the government about seven billion dollars per year—which is still cheaper than storing grain domestically. While the extreme drought in 1988 ameliorated some of the storage and subsidy problems, they are likely to return to haunt farm and export policies in the 1990s as the cycle swings once again from perceptions of scarcity to problems of abundance.

▼ Famine 1990

Turning to the famine side of the food market, there are dozens of countries that cannot now produce enough food to feed their rapidly growing populations but also cannot afford to buy any of the expensive American surplus. Moreover, most of these countries are not eligible for U.S. concessionary programs because they don't pass the political litmus test required of recipients of U.S. food aid. Of the countries in sub-Saharan Africa most affected by the drought in the 1980s, only two, the Sudan and Zaire, were privileged to be on the list of approved Public Law 480 beneficiaries.[55]

Although there are large pockets of famine and malnutrition throughout the less developed world, it is in Africa that the major Malthusian drama of the twenty-first century will be played out. A number of factors are contributing to the food disaster hovering over sub-Saharan Africa. Populations are growing at an average rate of 3.1 percent for the region, ranging from a high of more than 4 percent per year in Kenya to 1.7 percent annually in Gabon. Furthermore, these rapidly expanding populations are moving from the countryside to cities, thus leaving fewer people on the farms to meet burgeoning food needs. Urban populations are growing at an annual rate of between 6 and 8 percent in most of the countries of the region. Because of the demographic momentum inherent in the age structure of these populations, the current population of about 450 million is expected to grow to at least 1.4 billion by 2025. These problems are exacerbated by the approximate 2.7 million political, economic, and famine refugees in the region.[56]

Although only about one quarter of the potentially arable land in Africa is currently being farmed, there is little indication that rapid expansion of food production is likely. In fact, most of the region was self-sufficient in food during the early 1960s, but by the end of the 1980s it is estimated that food production per capita will have dropped by 20 percent. Only about 19 percent of the potentially arable land could be opened up without installing expensive irrigation systems, and about 45 percent of the potentially arable land is located in areas characterized by major endemic diseases. Furthermore, it is estimated that in the year 2000, if subsistence farming remains the norm in the region, twenty-one countries will exceed the human-population-carrying capacity of the land.[57]

Responsibility for much of the decline in sub-Saharan Africa has been attributed to a prolonged drought. But a close look at Africa's historical record indicates that the recent years are typical for the region and that the two decades prior to 1970 were very unusual in terms of rainfall.[58] It is likely that the increased nutritional needs of rapidly growing populations in the 1960s and early 1970s were masked by abnormally high levels of rainfall and resulting food production and that the starvation and malnutrition of the 1980s represent a delayed Malthusian adjustment to intolerable population pressures. Africa has historically been characterized by a nomadic way of life and large migrations of peoples. It could well be that this nomadic lifestyle has been a social evolutionary response to climatological irregularities and that the contemporary borders and dense populations thwart the traditional way of dealing with climate instabilities.

The African situation presents a serious challenge to many liberal values associated with food aid programs. Much to the dismay of many liberals, ecologist Garrett Hardin has suggested the practice of lifeboat ethics and triage in dealing with impending mass starvation.[59] He argues that famine is nature's response to population growth and that external food assistance, in the absence of effective population policies, makes little sense. It is instructive, in this regard, that of the twenty-three countries in sub-Saharan Africa, only five are rated as having even weak family planning programs.[60]

It is clear that complex and difficult choices must be made in dealing with the world's feast and famine problem. In an ideal world, Africa would develop enough purchasing power to rid the United States and Western Europe of food surpluses. But there are few indications that significant per capita economic growth will take place in Africa over the next two decades. In fact, sub-Saharan Africa has been running larger trade deficits over time, and there is little reason to expect a reversal of this trend in the near future. Beleaguered lending institutions aren't about to make new loans to many of these countries, which are seen as major credit risks. And, aside from the minuscule proceeds from antifamine rock concerts, additional food aid is unlikely to come from industrial countries already grappling with large budget deficits. Effective family planning programs are essential if this and other affected regions of the world are to move beyond heavy dependence on food imports toward some type of self-sufficiency in agriculture. Similarly, countries with food surpluses must develop mechanisms for dealing with them and for channeling excess production at affordable prices to countries that give evidence of resolving the long-term Malthusian problem.

▼ Notes

1. Lester Brown, "Reducing Hunger," in Lester Brown et al., *State of the World 1985* (Washington, D.C.: Worldwatch Institute, 1985), p. 29.
2. Figures taken from *Agricultural Outlook* (Washington, D.C.: U.S. Department of Agriculture, January/February 1982, January/February 1988).
3. See, for example, Georg Borgstrom, *The Hungry Planet* (New York: Collier Books, 1972); William Paddock and Paul Paddock, *Famine 1975!* (Boston: Little, Brown, 1967).
4. Lester Brown, op. cit., pp. 24–27.
5. Ibid., pp. 39–41.
6. See Reid Bryson and Thomas Murray, *Climates of Hunger* (Madison: University of Wisconsin Press, 1977); Stephen Schneider, *The Genesis Strategy* (New York: Plenum, 1976).
7. OECD, *World Supply and Demand of Major Agricultural Commodities* (Paris: 1976), pp. 29–30; Food and Agricultural Organization of the United Nations, *Assessment of World Food Situation* (Rome: 1974), p 36.
8. Paul Ehrlich et al., *Ecoscience: Population, Resources, Environment* (San Francisco: W. H. Freeman, 1977), App. 2.
9. Food and Agricultural Organization of the United Nations, op. cit., p. 36, and World Food Report 1986 (Rome: 1986), p. 65.
10. Paul Ehrlich et al., op. cit., pp. 974–978.
11. Lester Brown, *By Bread Alone* (New York: Praeger, 1974), pp. 23–25.
12. Cornelius Walford, "The Famines of the World: Past and Present," *Royal Statistical Society Journal* 41 (1878).
13. Cited in Paul Ehrlich and Anne Ehrlich, *Population, Resources, Environment* (San Francisco: W. H. Freeman, 1972), p. 13.
14. Sandra Postel, *Water: Rethinking Management in an Age of Scarcity* (Washington, D.C.: Worldwatch Institute, 1984), pp. 11–18.
15. Thomas Malthus, *An Essay on the Principle of Population as it Affects the Future Development of Society* (1798).
16. See Ralph Hardy, "Biotechnology: Status, Forecast and Issues," in Anne Keatley, ed., *Technological Frontiers and Foreign Relations* (Washington, D.C.: National Academy Press, 1985).
17. Reid Bryson, op. cit., chap. 6.
18. President's Science Advisory Committee, *The World Food Problem* (Washington, D.C.: 1967), vol. 2, p. 434.
19. OECD, op. cit., pp. 48–54.
20. *The Global 2000 Report to the President* (Washington, D.C.: U.S. Government Printing Office, 1980), p. 16.
21. Lester Brown, "Reducing Hunger," p. 25.
22. Ibid., p. 30; *The Global 2000 Report to the President,* vol. 3, p. 100.

23. *The Global 2000 Report to the President,* p. 13.
24. Ibid., vol. 3, p. 99.
25. See *World Resources 1986* (New York: Basic Books, 1986), pp. 264–265.
26. Figures on streamflow are from Cynthia Hunt and Robert Garrels, *Water: The Web of Life* (New York: Norton, 1972); see also Sandra Postel, op. cit., pp. 19–27.
27. See Paul Ehrlich and Anne Ehrlich, op. cit., p. 77.
28. See Sandra Postel, "Managing Freshwater Supplies," in Lester Brown et al., *State of the World 1985* (Washington, D.C.: Worldwatch Institute, 1985), pp. 50–56.
29. *The Global 2000 Report to the President,* op. cit., pp. 16–19.
30. Ibid., vol. 3, p. 101.
31. *World Resources 1986,* op. cit., pp. 264–265.
32. Lester Brown, "Reducing Hunger," op. cit., p. 31.
33. Data are from David Pimentel et al., "Food Production and the Energy Crisis," *Science* (November 2, 1973); Lester Brown and Sandra Postel, "Thresholds of Change," in Lester Brown, ed., *State of the World 1987* (Washington, D.C.: Worldwatch Institute, 1987), p. 11; Lester Brown, "Sustaining World Agriculture," in Lester Brown, ed., *State of the World 1987,* p. 131.
34. John Steinhart and Carol Steinhart, "Energy Use in the U.S. Food System," *Science* (April 19, 1974).
35. David Pimentel et al., op. cit.; see also David Pimentel and Marcia Pimentel, *Food, Energy and Society* (New York: Wiley, 1979); Maurice Green, *Eating Oil: Energy Use in Food Production* (Boulder, Colo.: Westview, 1978).
36. Lester Brown, *By Bread Alone,* op. cit., p. 117; Lester Brown, "Reducing Hunger," op. cit., p. 32.
37. See Ralph Hardy, op. cit. For a more critical view see Jack Doyle, *Altered Harvest: Agriculture, Genetics and the Fate of the World's Food Supply* (New York: Penguin, 1986).
38. Reid Bryson, op. cit., p. 154.
39. Stephen Schneider, op. cit., chap. 5.
40. See Richard Stothers, "The Great Tambora Eruption in 1815 and its Aftermath," *Science* (June 15, 1984).
41. Reid Bryson, op. cit., pp. 153–155.
42. See Richard Kerr, "Is the Greenhouse Here?" *Science* (February 5, 1988). See also the discussion of the greenhouse effect in Chapter 5.
43. See J. D. Hays et al., "Variations in the Earth's Orbit: Pacemaker of the Ice Ages," *Science* (December 10, 1976); Richard Kerr, "The Sun is Fading," *Science* (January 24, 1986); *Climate Change to the Year 2000* (Washington, D.C.: National Defense University, 1978).
44. These figures are estimates made in the mid 1970s and now are obviously on the conservative side. See Schlomo Reutlinger and Marcelo Selowsky, *Malnutrition and Poverty: Magnitude and Policy Options* (Baltimore: Johns Hopkins University Press, 1976).
45. Lester Brown, "Sustaining World Agriculture," op. cit., p. 134.
46. *Handbook of Economic Statistics 1987* (Washington, D.C.: Central Intelligence Agency, 1987), Table 157.
47. Ibid.
48. Bureau of the Census, *Statistical Abstract of the United States 1986* (Washington, D.C.: U.S. Department of Commerce, 1986), Table 1166.
49. See Barbara Insel, "A World Awash in Grain," *Foreign Affairs* (Spring, 1985).
50. *Foreign Agricultural Trade of the United States: Calendar Year 1986 Supplement* (Washington, D.C.: U.S. Department of Agriculture, 1987), p. 30.
51. Barbara Insel, op. cit., p. 895.
52. Ibid., p. 898.
53. "The Good Earth is Bad News," *Business Week* (August 19, 1985).

54. "Hard Times Will Get Harder Down on the Farm," *Business Week* (January 13, 1986).
55. Barbara Insel, op. cit., p. 897.
56. See Thomas Goliber, "Sub-Saharan Africa: Population Pressures on Development," *Population Bulletin* (February 1985).
57. See John Walsh, "The Return of the Locust: A Cloud over Africa," *Science* (October 5, 1986); C. M. Higgins et al., *Potential Population Supporting Capacities of Lands in the Developing World* (Rome: Food and Agricultural Organization of the United Nations, 1982), pp. 31, 137.
58. See National Research Council, *Environmental Change in the West African Sahel* (Washington, D.C.: National Academy of Sciences, 1983); *Food Problems and Prospects in Sub-Saharan Africa* (Washington, D.C.: U.S. Department of Agriculture, 1981); Lester Brown and Edward Wolf, *Reversing Africa's Decline* (Washington, D.C.: Worldwatch Institute, 1985).
59. Garrett Hardin, "Living in a Lifeboat," *Bioscience* (October 1974); Garrett Hardin, *The Limits of Altruism* (Bloomington: Indiana University Press, 1977); George Lucas and Thomas Ogletree, eds., *Lifeboat Ethics: The Moral Dilemmas of World Hunger* (New York: Harper & Row, 1976).
60. Thomas Goliber, op. cit., p. 39.

▼ Suggested Reading

LESTER BROWN AND EDWARD WOLF, *Reversing Africa's Decline* (Washington, D.C.: Worldwatch Institute, 1985).

PETER BROWN AND HENRY SHUE, eds., *Food Policy: The Responsibility of the United States in the Life and Death Choices* (New York: Macmillan, 1978).

REID BRYSON AND THOMAS MURRAY, *Climates of Hunger* (Madison: University of Wisconsin Press, 1977).

MICHAEL DOVER AND LEE TALBOT, *To Feed the Earth* (Washington: World Resources Institute, 1987).

JACK DOYLE, *Altered Harvest: Agriculture, Genetics and the Fate of the World's Food Supply* (New York: Penguin, 1986).

Food Problems and Prospects in Sub-Saharan Africa (Washington, D.C.: U.S. Department of Agriculture, 1981).

SUSAN GEORGE, *How the Other Half Dies: The Real Reasons for World Hunger* (Montclair, N.J.: Allenheld, Osmun and Company, 1977).

MAURICE GREEN, *Eating Oil: Energy Use in Food Production* (Boulder, Colo.: Westview, 1978).

GARRETT HARDIN, *The Limits of Altruism* (Bloomington: Indiana University Press, 1977).

W. LADD HOLLIST AND F. LaMOND TULLIS, eds., *Pursuing Food Security: Strategies and Obstacles in Africa, Asia, Latin America, and the Middle East* (Boulder, Colo.: Lynne Rienner, 1987).

MITCHELL KELLMAN, *World Hunger: A Neo-Malthusian Perspective* (New York: Praeger, 1987).

FRANCES LAPPE AND JOSEPH COLLINS, *World Hunger: Twelve Myths* (New York: Grove Press, 1986).

GEORGE KENT, *The Political Economy of Hunger* (New York: Praeger, 1984).

ALEX MCCALLA AND TIMOTHY JOSLING, *Agricultural Policies and World Markets* (New York: Macmillan, 1985).

JOHN MELLOR, CHRISTOPHER DELGADO, AND MALCOLM BLACKIE, eds., *Accelerating Food Production in Subsaharan Africa* (Baltimore: Johns Hopkins University Press, 1987).

WILLIAM MURDOCH, *The Poverty of Nations: The Political Economy of Hunger and Population* (Baltimore, Md.: Johns Hopkins University Press, 1980).

National Policies and Agricultural Trade (Paris: OECD, 1987).

L. A. PAULINO, *Food in the Third World: Trends and Projections to 2000* (Washington, D.C.: International Food Policy Research Institute, 1986).

ROBERT PAARLBERG, *Fixing Farm Trade: Policy Options for the United States* (Cambridge, Mass.: Ballinger, 1987).

DAVID PIMENTEL AND MARCIA PIMENTEL, *Food, Energy and Society* (New York: John Wiley, 1979).

ALAIN REVEL AND CHRISTOPHE RIBOUD, *American Green Power* (Baltimore, Md.: Johns Hopkins University Press, 1987).

SANDRA POSTEL, *Water: Rethinking Management in an Age of Scarcity* (Washington, D.C.: Worldwatch Institute, 1984).

STEPHEN SCHNEIDER, *The Genesis Strategy* (New York: Plenum Press, 1976).

RADHA SINHA, ed., *The World Food Problem: Consensus and Conflict* (Oxford: Pergamon Press, 1978).

ALBERTO VALDES, *Food Security for Developing Countries* (Boulder, Colo.: Westview Press, 1981).

JOHN WARNOCK, *The Politics of Hunger: The Global Food System* (New York: Methuen, 1987).

5

Resources, Commons, and National Security

Energy from food and fossil fuels is of critical importance for human well-being, but a variety of other resources are also crucial for economic prosperity and national security. Demand for conventional resources, such as nonfuel minerals, has grown apace with the spread of the industrial revolution. These materials are in many ways similar to petroleum and food. They are traded on international markets, are subject to boom and bust price cycles, and most of the industrial countries import substantial quantities of them from the less developed world. Other kinds of natural resources are also very important to future human welfare but are not traded on world markets. These nonconventional resources, such as air and water, are vital to all human beings, but since they are used by all mankind in common, they are not always thought of as valuable. If current trends continue, however, it may well be that problems with water and air become threats to economic and physical well-being long before the world runs out of the more conventional resources.

The adequacy of future supplies of nonfuel minerals has been the subject of considerable controversy in debates between technological optimists and transformationists. On the one hand, an argument can be made from an ecological point of view that the depletion problems that bedevil petroleum are also characteristic of many nonfuel minerals.[1] On the other hand, arguments from an economic perspective stress the role of technology and higher prices in bringing new but perhaps less rich reserves into production.[2] Like petroleum, these minerals are often produced in large quantities by less developed countries and imported by many industrial countries. As mineral dependencies have grown, there has been considerable speculation about the possibility for exporters of key minerals to form exporter cartels along OPEC lines and thus exploit these growing vulnerabilities.[3]

Table 5-1 indicates levels of import dependence for ten key industrial minerals. Despite efforts to attain more self-sufficiency, Japan has been and remains highly dependent on other countries. While the European Economic Community is in somewhat better shape, over half of its consumption of these ten key minerals is supported by imports. As is the case with petroleum, the United States is comparatively better off, although it imports increasingly large quantities of minerals from less developed countries.

Table 5-1 Import dependence for selected minerals (net imports as percentage of consumption).

	United States		EEC		Japan	
	1972	1987	1972	1986	1972	1986
Bauxite and alumina	88	97	51	52	100	100
Copper	17	25	93	98	90	88
Iron ore	32	28	37	94	94	99
Lead	19	15	75	84	76	66
Nickel	90	74	89	41	100	100
Zinc	55	69	61	63	80	61
Tin	100	73	96	56	97	96
Tungsten	42	80	100	46	100	73
Manganese	95	100	98	98	90	100
Chromium	100	75	100	96	100	98

SOURCE: U.S. Bureau of Mines.

Aside from the more general balance-of-trade considerations associated with significant mineral dependence, there is also a set of more immediate national security considerations related to certain nonfuel minerals. A subset of these minerals, the strategic materials, is found only in a few less developed countries, often only in southern Africa. Political uncertainties in that part of the world have heightened concerns about the future availability of these minerals, many of which are critical components of high-technology military equipment. One of the spillover effects of the OPEC decade is a feeling of urgency in some industrial countries that supplies of these strategic materials as well as other nonfuel minerals will be interdicted, either by new cartels or political-military events. Task forces have been set up to evaluate mineral vulnerabilities, reports have been done on substitutes for strategic minerals, and stockpiles have been set up in anticipation of the tough times that are thought to be ahead.[4]

The nonconventional resources have received less public concern and attention, largely because they are less visible problems of the global commons and their immediate relationship to national security is less obvious. The air in the world's atmosphere and the water in the hydrosphere are constantly in motion and do not respect national boundaries. Unlike more conventional resources, the problem is not that these resources will be used up in the near future, but rather that their composition will be altered and motions changed by pollution, thus bringing on a host of climatological problems. As the world's population and living standards have steadily grown, larger quantities of pollutants—the by-products of prosperity—have been put into the air, streams, and oceans. Although it is common to say that these wastes have been disposed of in the oceans or the atmosphere, in reality most pollutants have been dispersed, or reduced in concentration to levels that are less toxic and noticeable.

Traditional ways of handling pollutants are no longer adequate in a much more densely populated and industrialized world, because the atmospheric and aquatic resources that have been used to disperse pollutants can no longer carry the burden. One of the first environmental features seen by early astronauts orbiting the earth was a long plume of exhaust gases stretching from the East Coast of the United States into Western Europe, where the plume was augmented by Western European pollution. Studies of the world's oceans reveal that the surface is covered with bits of plastic as well as a thin layer of petroleum.[5] Many of the world's aquifers, underground water supplies, are now polluted by surface runoff and soon will not be usable for drinking water.[6] In summary, when human numbers and demands on these environmental waste dispersal systems were small, the air and water were cheap ways of moving potentially dangerous wastes. But with growing populations and increasing worldwide industrialization, the atmosphere and hydrosphere are being taxed beyond capacity, and these resources could also be considered to be in danger of depletion.

Apart from the more obvious problems associated with local and regional pollution of air and water, two major global issues and a set of transborder pollution problems are now high on the international political agenda. In the former category are the buildup of carbon dioxide in the atmosphere and the depletion of the earth's ozone layer by chlorofluorocarbons and their close relatives. In both cases, industrial country production and transportation systems are contributing to potential catastrophes of the global commons, but only limited collective action has been taken to address the potential crises. In the latter category, the transport of acid rain across national borders has been observed and well documented, but only hesitant remedial action has yet been taken. Thus, although the integrity of the atmosphere and hydrosphere is not usually thought of as a security problem, the political and economic costs associated with damage done to it might well eventually be much higher than with more conventional types of mineral dislocations.

▼ Minerals and Markets

Assuring adequate and affordable flows of conventional resources into the international political economy remains a major security concern. Leaders of many mineral-exporting less developed countries see a potential opportunity to mimic the OPEC successes of the 1970s by forming producer associations to control exports of basic commodities. Along with the redistribution of power that accompanied the early OPEC successes, the Group of 77 (now numbering many more than one hundred) less developed countries presented a manifesto to the Sixth Special Session of the General Assembly of the United Nations in 1974 demanding the establishment of a New International Economic Order. Among the many issues covered by the document was a proposal for international cooperation in stabilizing the prices of a long list of basic commodities through producer–consumer agreements.[7] Lacking support from the industrial countries, the proposed agreements for the most part never came into force; thus the less developed countries were left with the option of trying to form producer associations without the cooperation of importing countries.

Buoyed by the heady atmosphere of rapid economic growth and rising commodity prices in the early 1970s, exporters of bauxite and copper attempted, with some success, to increase their control over prices of these commodities.[8] And these initial successes led to speculation that similar efforts in other commodity markets were likely.[9] Subsequent experience has shown that such efforts face serious economic obstacles, and although various experiments have produced some positive short-term changes for some countries involved, there have been no clear persisting benefits.

It is pointless to speculate about cartels, future prices, and potential interdictions of supplies without knowing something about the physical nature of each mineral, the economics of its extraction and marketing, and the uses for it. Nonfuel minerals and their markets are complex, and it is difficult to generalize about their characteristics. Some of them are much more important economically than are others. Iron ore, bauxite, and copper, for example, could be called the building blocks of industrial civilization, whereas the ferro-alloys—minerals that are alloyed with iron to make specialty steels—have greater strategic importance. In contrast, the world is not likely to collapse if there is no lead or zinc. Minerals also vary a great deal in terms of worldwide proved reserves in relation to current consumption. There is little reason to fear running out of iron ore in the next century, for example, but there is concern about copper and chromium.[10] Finally, the number of countries with reserves and their location in relation to future markets is also important. World reserves of iron ore are widely distributed, but chromium ore is found mostly in southern Africa and the Soviet Union. In summary, some minerals are more important and therefore worthy of greater attention, whereas others are intrinsically interesting only because they are representative of certain categories. It is neither feasible nor desirable to analyze the future prospects for dozens of minerals here, and only a few are examined.

Iron ore is representative of a large group of minerals for which there is little potential for successful exporter collusion in the foreseeable future. In quantity consumed, iron ore is the most significant nonfuel mineral, and historically it has been the backbone of the industrial revolution. Iron was used to make implements long before 1000 B.C., but it was not until the invention of the blast furnace at the beginning of the eighteenth century that iron—and eventually steel—could be produced in large quantities. Blast furnaces smelt iron ore into pig iron, which in turn is transformed into cast iron, wrought iron, or various kinds of steel. Cast iron is used to make machine parts and has many other industrial applications. Wrought iron is softer and more resistant to corrosion and is often used for pipes and for ornamental work. Most iron ore, however, is used to make various kinds of steel. This is done by alloying iron with various minerals in order to obtain steels with desired characteristics. Among the principal alloys used in making specialty steels are manganese, chromium, nickel, tungsten, vanadium, and molybdenum. The most valued attribute of steel is its strength. It is essential in building construction, and it is used to reinforce concrete, in the holds of ships, and in mass transit construction. Thousands of other specialty items—including defense equipment—are produced from the many alloys.

Because of its central role in transportation, defense, and heavy construction, it is not realistic to speculate about developing substitutes for iron. But there is little reason to worry about the issue, because a number of characteristics of this mineral and its market make it an unlikely candidate for cartelization. Iron ore is one of the most abundant elements in the earth's crust, and the present cutoff grade (the richness of the ore that is now being economically mined) is only 3.4 times the average crustal abundance of the mineral.[11] Ecological scarcity doesn't enter the picture either, because it is projected that proven reserves of iron ore will be adequate to meet demands for at least two hundred years.[12] In addition to these reserves, more than one third of the iron consumed around the world now comes from recycled scrap. Furthermore, there are several major producers of iron ore throughout the world, and nine countries contribute more than 5 percent of world production.

In 1975, eleven countries signed an agreement giving birth to a producer association called the Association of Iron Ore Exporting Countries (AIOEC). The signatories of the agreement were Algeria, Australia, Brazil, Chile, India, Mauretania, Peru, Sierra Leone, Sweden, Tunisia, and Venezuela. It is difficult to imagine a politically and economically more diverse group of countries. The agreement does not give the association any price-fixing powers, and the very diverse participants are bound only to consult with and mutually aid one another.[13]

Iron ore is typical of a number of minerals that are produced in many countries and that are abundant in relation to demand for them. But there are potential economic and security problems in the American iron and steel industry that, in some ways, parallel those in the petroleum industry. The United States has been steadily losing jobs and production in the steel industry to foreign competition. This is a result of richer ore deposits that are less costly to mine, newer and more efficient plants and equipment, and lower wages paid in other countries. Thus, the key problem for the United States is a developing vulnerability caused by the migration of the iron and steel industry to low-cost countries rather than any outright disappearance of the mineral.

The world aluminum industry is somewhat similar to the iron and steel industry, but there are important differences. Aluminum is second only to iron in terms of quantity of ore consumed. It is an important building block of the latter stages of the industrial revolution and is used extensively in the construction industry when the rigidity of steel is not necessary. It is also important in the aircraft and automobile industries, where its light weight and durability are important. Aluminum ore is spread abundantly in the earth's crust and is found in many different clays and oxides. Bauxite is the type of aluminum ore most widely mined. The present cutoff grade for the ore is only 2.2 times the average abudance of it in the earth's crust.[14] In addition, new deposits of ore are constantly being found and proven reserves have increased over 300 percent in the last twenty-five years.[15] Ore-bearing clays that will become potential reserves should the price of bauxite double are also abundant.

The world aluminum market is unique because it is tightly controlled by a handful of vertically integrated multinational corporations. Six large companies—Alcoa, Alcan, Reynolds, Kaiser, PUK (France), and Alusuisse—control over 50 percent of the aluminum capacity in the market economies.[16] Accurate data on ore production, reserves, and prices are not easily available, since most transactions take place within these companies. Only the price of finished aluminum is normally quoted by manufacturers. Thus, in the case of bauxite, the producer–consumer dialogue takes place between producer countries and multinational corporations, often a test of wills between developing countries and corporate giants. Aluminum prices are also affected by the energy-intensive nature of production, the price of aluminum being determined less by the cost of the ore—which has, in recent years, made up between 7 and 12 percent of the final aluminum price—and more by the cost of electricity used in aluminum production.

Seven major bauxite-producing countries, accounting for 65 percent of world production and 80 percent of world trade, formed the International Bauxite Association (IBA) in 1974. They were later joined by four other countries, bringing current membership to eleven. Ten of the members are less developed countries; Australia is the only industrialized exporter. On the surface it would seem that the aluminum market would be ripe for producer collusion. However, most bauxite is traded via internal transfers within multinationals: this control of downstream facilities dampens the potential for concerted producer action to raise prices. Futhermore, bauxite prices suffer less from price instability than other commodities, largely because of the ability of producing countries to negotiate stable returns with the multinational corporations that control the industry.

Individual members of the IBA have attempted to increase unilaterally the revenue gained from bauxite exports, but they have not been able to get other countries to follow their lead. Jamaica increased royalties and export levies on bauxite shipped to American companies by about 150 percent in 1974–75. Within a year five other members of the IBA had similarly raised prices. It was possible for these countries to gain more revenue from exports because some production facilities were designed only to accommodate certain types of ore. In the short term, the companies were willing to shoulder the higher prices rather than retool production processes to accommodate different ores. But in the long run, there are severe limits on upward price pressures because of the expansion of mining operations in Australia and Guinea and the development of reserves in Brazil.

The multinational corporations making aluminum in the United States have been importing between 85 and 90 percent of the bauxite used in the production process. Jamaica has supplied a large percentage of the ore, with smaller amounts coming from Suriname, Guyana, and Australia. Despite this high degree of dependence, there is little reason to worry about future U.S. vulnerability. The multinationals that control the downstream facilities seem capable of dealing effectively with producer countries and can shift purchases from one country to another when necessary, particularly when markets are oversupplied. Since the big aluminum companies control the energy-intensive downstream facilities, even a doubling of bauxite prices would add only a small sum to the final price of aluminum. From a U.S. national security perspective, there is a significant stockpile of aluminum to be used in case of emergency, and a variety of more expensive domestic clays can also be used should companies choose to exploit them.

Copper is a "transindustrial" mineral used in many industrial and post-industrial applications, and there are some legitimate security concerns about its future availability. It has a number of important properties, such as ductility, conductivity, resistance to corrosion, and a high melting point, that make substitutions for it difficult. About 80 percent of the copper consumed in the United States is used in the electronics and communications industries. Although aluminum and steel wires have been used with mixed success in high-power transmission lines, and fiber-optic cable promises to be a substitute in many communications processes, there are now no practical substitutes for copper in electric motors, where heat is an important limiting factor.[17]

Unlike iron ore and bauxite, copper is not found abundantly or randomly throughout the earth's crust. There is currently enough copper in proved reserves to last between twenty and forty years, given varying assumptions about increased demands for it.[18] Although a number of countries produce copper, the export market is dominated by four less developed countries—Chile, Peru, Zambia, and Zaire. Because of the relatively tight supply situation and perceptions of ecological scarcity, during periods of economic expansion in the 1970s the price of copper rose to well over $1.25 per pound but fell to $.65 per pound during the commodity bust of the mid 1980s before recovering substantially at the end of the decade. Copper does not exist in unlimited exploitable quantities; since the world export market is dominated by four less developed countries, it would seem to be an ideal candidate for producer collusion.

The Council of Copper Exporting Countries (CIPEC) was formed by the four major LDC exporters in 1967, in response to a substantial decline in copper prices during the previous year. In the early years, governments of these countries directed their efforts at nationalization of mines owned by multinational corporations, and for the most part they succeeded. In the wake of the initial OPEC successes and a sharp drop in copper prices in 1974, CIPEC countries agreed to reduce copper output by 15 percent in 1975. But political and economic realities soon undermined the price stabilization effort. Following the demise of the Allende regime and the rise of a market-oriented military dictatorship in Chile, the government officially ended the production cuts in May 1976. In addition, Papua New Guinea and the Philippines emerged as low-cost producers outside of the original agreement. In light of these events and the fact that earnings from copper exports accounted for more than 50 percent of the export earnings of Zambia, Zaire, and Chile, the efforts to influence copper prices were doomed to failure.[19] The low-cost exporters, Chile, the Philippines, and Papua New Guinea, have large reserves and seek expanded markets. They have little interest in participating in schemes that would mostly benefit Zaire and Zambia, the high-cost producers. Furthermore, major price increases would spur exploration in other countries as well as hasten the development of seabed mining, potentially opening up billions of tons of new copper reserves.

The United States has significant copper reserves and meets only a small portion of its needs through imports. But the copper industry in the United States is in trouble as much production has migrated to other countries. In the early 1980s, imports of copper into the United States more than doubled, and the trend toward more imports has continued. Nearly two thirds of these imports come from Chile, which claims that rich reserves combined with low wages permit production of copper at only forty-eight cents per pound, well below the break-even cost of producing it in the United States, which is about eighty cents per pound. In the mid 1980s, some 40 percent of the copper mining capacity of the United States was shut down, and the industry sought relief from cheap imports through the U.S. International Trade Commission. Chile, on the other hand, claims that copper exports into the U.S. market are essential if the massive debts owed to U.S. banks are to be serviced.[20]

Remaining nonfuel minerals can be divided into three categories: the precious metals, ferro-alloys, and nonferrous metals. The precious metals are familiar and include gold, silver, and platinum. They have historically been valued for their luster and have been used mainly in jewelry, although they also have been used as a hedge against inflation. Aside from their psychological value, these metals are increasingly being used in industry. Silver is used in the electronics industry, and platinum is used as a catalyst in the petrochemical industry as well as in catalytic converters in automobiles.

The nonferrous metals, of which copper is the principal one in international trade, also include lead, zinc, and tin. World supplies of lead are more than adequate to meet projected demand, and its low melting point means that large quantities can be recycled easily and cheaply. Zinc, used in galvanizing and die-casting processes as well as in photography, is another matter. Because of a lack of world reserves, zinc may soon disappear as a widely used metal. No high grade deposits of zinc have been discovered in the United States in decades, and dependence on foreign sources has been rising over the years.

Perhaps the most interesting of these otherwise homely and proletarian metals is tin. Tin is used to make pewter and bronze, as solder in computers and other electrical equipment, and as a coating for the steel in tin cans, which contain a lot of steel and very little tin. But the tin market is most interesting because it is one in which an agreement between producers and consumers stabilized prices over a thirty-year period beginning in 1955. The International Tin Agreement has been looked on as a prototype agreement between producers and consumers because it utilizes a buffer stock to keep tin prices within an agreed target range on the world market. The tin agreement is a series of five-year agreements that have established an International Tin Council to supervise and administer them. The council attempts to stabilize tin prices by buying and selling from the buffer stock, by asking members to use export controls for short periods, and by indicating optimal price ranges in case the buffer stock proves inadequate.[21]

During the course of the International Tin Agreement, rising demand for tin kept the price well above the agreed floor level, and on several occasions the price pushed through the ceiling. But the tin bubble burst in 1986, when a protracted global commodity slump wreaked havoc. In seeking to support the floor price, the International Tin Council enlarged the buffer stock to clear a glutted market. But the council ran out of money with contractual obligations to purchase eighty thousand tons of tin. The price of tin dropped from $13,080 to $5,880 per metric ton in the ensuing panic, and it is unclear how and when this most durable of producer–consumer agreements can be put back together.[22]

▼ Strategic Materials

The possibility of mineral exporting countries wreaking economic havoc in industrial countries by withholding exports is now very low, particularly in the absence of a return to rapid global industrial expansion. But this doesn't mean that imported minerals are no longer a strategic concern for the United States. Certain minerals have important defense applications, are imported by the United States in significant quantities, and come from a very small number of countries. Many of these critical minerals are ferro-alloys, metals that are alloyed with iron in order to produce various kinds of specialty steels. Often they are added in only very minute quantities, sometimes making up as little as 1 percent of the final product, to yield desired properties such as hardness, luster, durability, or resistance to high temperatures. The major ferro-alloys are chromium, manganese, cobalt, vanadium, nickel, molybdenum, niobium, and tungsten. These and other critical minerals have different unique applications, and there are no readily available substitutes for them.

A second important strategic concern for the United States is the massive natural resource treasury found in the Soviet Union. Figure 5-1 illustrates the strikingly different positions of the United States and the Soviet Union. The USSR is resource-rich, being self-sufficient or nearly self-sufficient in twenty-eight of thirty-four important nonfuel minerals. Even though this imbalance has no immediate geopolitical consequences, it is obvious that the Soviet Union would be much less affected by dislocations in mineral markets than would the United States, and it might even be able to profit from them.

A third strategic issue of concern for the United States is the migration of minerals industries from the United States into the Third World. There are two reasons for this migration. First, there are large wage differences in the world mining industry, and foreign production is much cheaper. Second, although reserves of various mineral ores in the United States could be exploited, there are richer and cheaper reserves in many of the less developed countries. The bottom line is that low-wage foreign producers operating with richer ore bodies have been overwhelming the U.S. mining industry. In the period from 1979 to 1983, for example, jobs in metals mining dropped from 109,000 to 44,800, and they have declined more slowly since then.[23]

Finally, several of these critical minerals have been identified as being so important to the security of the United States, so limited in supply, and located in such volatile areas of the world that they are classified as strategic materials. Figure 5-2 displays the thirteen minerals in this category with the regions that are currently major sources of supply. All of these minerals are imported in large quantities by the United States. Four of them—chromium, cobalt, manganese, and the platinum metals group—are so important to the U.S. economy as to be designated first-tier strategic materials.[24]

Chromium is the most critical of these first-tier strategic materials and occasions the greatest security concern. It has a variety of applications throughout the economy. In its mineral form, chromite, it is used in boiler fireboxes, furnaces, and in foundry sands used for casting molds. Its most important uses, however, are as an alloying element, making steel harder and increasing resistance to wear. When combined with nickel, cobalt, aluminum, or titanium, chromium produces superalloys, which are used in high-temperature situations where resistance to friction is also required. Chromium ore is found in many parts of the world, but South Africa and the Soviet Union account for nearly two thirds of world production. In 1982 nearly half of total U.S. chromium imports were from South Africa, and significant quantities came from Zimbabwe, the Philippines, and Yugoslavia.[25] Western Europe and Japan also meet nearly 100 percent of their chromium needs from abroad and are similarly vulnerable to supply interruption. Of all of the nonfuel minerals, the chromium supply is thus most sensitive to political disruption, and efforts are under way to find methods of reducing this significant vulnerability.

The United States imports much less cobalt than chromium, but cobalt is also strategically important because it is essential in many of its applications. Cobalt is often mixed with iron to make permanent magnets. It is also a critical additive in the manufacture of some superalloys and is used as a binder in tungsten carbide drill

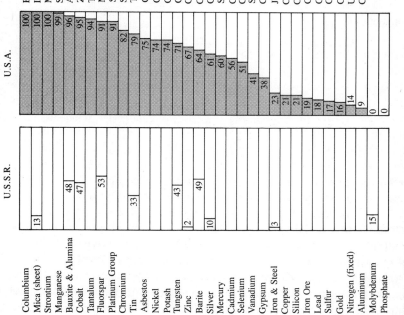

Figure 5-1 Net import reliance, 1984: selected nonfuel mineral materials.

SOURCE: U.S. Bureau of Mines.

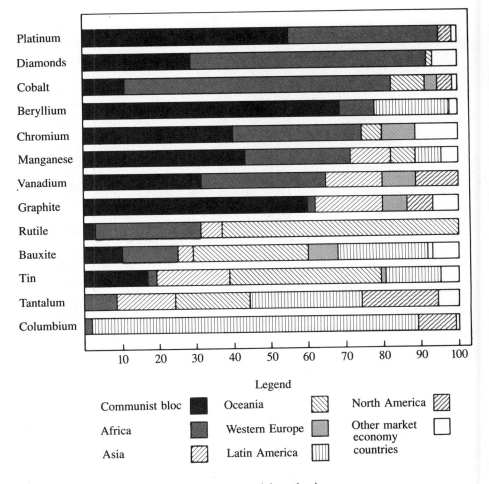

Figure 5-2 Regional distribution of strategic material production.
SOURCE: Office of Technology Assessment from U.S. Bureau of Mines data.

bits. Cobalt is also used in certain petroleum refining processes.[26] The world's supply of cobalt, like that of chromium, is concentrated in the southern part of Africa. Zaire produces about one half of the world's total, and neighboring Zambia produces more than 10 percent. Remaining cobalt production is divided among a number of small producers. The importance of the mineral combined with dominance of the industry by Zaire and Zambia have made it vulnerable to rapid price fluctuations. In 1978, for example, the price shot up from $5 to $50 per pound because of political unrest in Zaire. In 1986, the price of cobalt nearly doubled, from $3.60 to $7 a pound, because so much cobalt is exported through politically unstable South African ports.[27]

Manganese is used mainly as an alloy and processing agent in steel production. In the future, if the U.S. steel industry survives, large quantities will be imported. There is a much greater diversity of manganese exporters than is true for chromium or cobalt, but the Soviet Union and South Africa together control nearly two thirds of world production. Production of platinum group metals, which are important in petroleum refining, automotive exhaust systems, and telecommunications systems, is also concentrated in these two countries, with nearly all of the world's supply originating there.

The strategic minerals problem for the United States and several other industrial countries is exacerbated by the political situation in the southern part of Africa. This part of the African continent produces much of the world's supply of critical minerals, and in many cases the Soviet Union is the only major alternative supplier. South Africa itself possesses major reserves of most industrial minerals and is the world's fourth largest producer of them. It is a major supplier of first-tier critical minerals to the United States and other OECD countries. South Africa also dominates the economies and transportation networks in the region, possibly compounding future access problems.

The mineral wealth of South Africa and its strategic location in the mineral-rich portion of the continent have undoubtedly influenced U.S. policy toward apartheid. Interdiction of supplies of strategic minerals from South Africa would be economically disruptive to OECD countries and would strengthen the position of the Soviet Union, an alternative supplier of many of the minerals that would be in short supply. Probably the largest windfall for the Soviet Union would come from a disruption of South African gold exports, since South Africa and the Soviet Union are the world's major gold exporters.[28] In order to counteract the short-term impact of potential political disturbances in South Africa or other parts of the world, the United States now stockpiles about ten billion dollars' worth of materials in a national defense stockpile and is searching for substitutes for some of the most critical materials.[29]

▼ Nonconventional Resources and Ecological Interdependence

Most studies of the national security implications of resource vulnerabilities have focused nearly exclusively on short-term scarcities of energy, food, and strategic minerals and have almost totally ignored other essential resources. Without access to these strategic materials, defense industries might well grind to a halt until substitutes could be developed or military action taken to restore supplies, and strategic attention has been focused on tangible minerals and agricultural commodities, things that are owned and exported by various countries.

Much less attention has been paid to nonconventional—but equally critical—commonly used natural resources, such as air and water. These resources, although extremely important to the well-being of future generations in all nations, are less tangible and their linkage to national security considerations seems at first to be much more tenuous. It requires a certain amount of analytic skill and knowledge to understand their importance and the long-term threats to them. Worldwide abuse of the atmosphere and hydrosphere threatens to alter delicate balances in the global ecosystem in ways that might soon present acute threats to human health and economic well-being.

These nonconventional resources suffer from the fact that they are not claimed by individual nations but rather are the common property of the human race. Since no one claims ownership and the use of them is free, they are treated with a considerable amount of contempt, the assumption being that the supply of them will always be adequate to meet demand.

Treatment of these commonly used resources is well described by Garrett Hardin's classic tragedy of the commons paradox. He likens commonly held resources to a medieval pasture open to use by all. It seems rational to each herdsman using the pasture to graze as many animals as possible. The herdsman calculates that each additional animal munching on the pasture represents his personal gain. The problem, however, is that a large number of herdsmen acting in this same "rational" manner can overcrowd the pasture and lead to devastation of all herds through overgrazing. Thus, in the absence of either a superordinate authority to oversee the pasture or its privatization, a tragic deterioration of the land takes place.[30]

The leap from Hardin's pastoral commons to the contemporary global commons is a short one. The global atmosphere and hydrosphere are really two gigantic resource commons: the former supplies the countries of the world with fresh air and the latter is responsible for the global circulation of fresh water and thus world agricultural productivity. All nations share these collective resources, which are at the root of growing ecological interdependence, and each nation depends on the others to help preserve their integrity. But these commons often are seen by the individuals, corporations, and nations using them as a free resource, and they are heavily exploited by contemporary versions of herdsmen attempting to maximize the immediate production that can be wrung from them. In the case of the atmosphere, industrial and industrializing countries treat the world's supply of air as a gigantic waste dispersal system, gradually changing the composition of the atmosphere in ways that will eventually have dramatic consequences for all nations. In the case of the hydrosphere, not only is it similarly treated as a gigantic global septic tank for dispersing waste, but the oceanic portions of it are also under increasing pressure from fishing nations, which could lead to the collapse of world fisheries. At present, no international political institutions seem capable of regulating the health of these global commons.

The world's atmosphere consists of 21 percent oxygen, 78 percent nitrogen, and a host of other gases found in very small quantities. This delicate balance of gases circulating around the planet sustains all forms of life. The atmosphere not only provides oxygen to human beings and nitrogen to plants, but it also serves as an insulating blanket, keeping temperatures within a tolerable range and protecting life from the harsh effects of ultraviolet radiation from the sun. Unfortunately, even very small changes in the composition of the atmosphere can trigger large-scale changes in temperature and rainfall.

Recent studies indicate that the global herdsmen now profiting by despoiling the atmosphere may be altering the composition of it in ways that will shortly have a very negative impact on human well-being. First, although carbon dioxide is found in only minute quantities in the atmosphere, about 340 parts per million, data indicate that the amount of it is steadily increasing as a result of fossil fuel combustion and deforestation. Carbon dioxide plays an important role in regulating the earth's temperature, and this observed buildup of it is gradually increasing temperatures worldwide through a greenhouse effect. Second, although ozone makes up only .01 parts per million of the atmosphere, an ozone layer high up in it filters out dangerous ultraviolet radiation. There is considerable evidence that the atmospheric buildup of certain chlorofluorocarbons and related gases may be chemically destroying this protective ozone layer. Finally, sulphur and nitrogen oxides and related pollutants are creating significant transborder pollution problems that are becoming major irritants in relations among neighboring countries.

Flushing wastes into the hydrosphere is changing it in ways that are somewhat less dramatic, but the integrity of the world's constantly moving supply of water is also being threatened by this treatment. While 71 percent of the earth's surface is composed of water, only 1 percent of this water is available for human consumption. Salt water in the oceans makes up 97 percent of it, and another 2 percent is locked in the polar ice caps. Thus, the small percentage of water that is available in streams, lakes, and underground aquifers is heavily used by the world's growing population. In the United States, for example, an average drop of Mississippi River water making its way from Minneapolis to the Gulf of Mexico gets withdrawn from the river and used at least six times, resulting in drinking water in New Orleans laced with various kinds of toxic substances. Almost all of the world's long rivers have similar problems, as the competing demands of people, agriculture, and industry serve to diminish the purity of this water. Even many of the underground aquifers in the United States are now heavily polluted by agricultural chemicals that make their way into the soil along with rainfall.[31]

Many water problems are local, but the cumulative impact of polluted streams flowing into the world's oceans combined with direct mistreatment of them could well be causing small changes in the hydrosphere, which in turn are capable of triggering large-scale climatic shifts. Much of the world's ocean surface, for example, is coated with plastic particles and hydrocarbon molecules. This thin layer could inhibit evaporation of ocean water and interfere with normal movement of moisture through the hydrosphere, which in turn could alter the equilibrium of the system and change rainfall and temperature patterns over large stretches of the planet. Although some progress in regulating water pollution has been made on a regional level, no international institutions are currently capable of controlling pollution of the oceanic commons.[32]

▼ A Greenhouse Warming

The most significant long-term problem of ecological interdependence stemming from mistreatment of the atmospheric commons is the inexorable buildup of carbon dioxide gas in it. Carbon dioxide is a natural part of the atmosphere that is found in it in very small quantities. It is estimated that carbon dioxide made up about 280 parts per million of the atmosphere prior to the global spread of industrialization in the twentieth century. Over the last century, human activities have liberated significant amounts of carbon dioxide, and there are now about 340 parts per million. Estimates are that this figure will nearly double over the next fifty to eighty years.[33]

Carbon dioxide in the atmosphere has no direct adverse effects on human health. But these small quantities of the gas do have a significant impact on world temperatures and rainfall patterns. Carbon dioxide and other trace gases operate as a one-way screen in the atmosphere. They readily transmit most of the spectrum of radiation coming from the sun, but they absorb the deep infrared radiation given off by the earth. As the amount of carbon dioxide in the atmosphere continues to grow, radiation from the sun is still not much inhibited as it passes through the atmosphere on its way to the earth, but a growing portion of the infrared radiation coming from the earth cannot escape into outer space. Thus, the carbon dioxide acts very much like a greenhouse, permitting most solar radiation to pass through but trapping the infrared radiation from the surface, thus causing a long-term warming of the earth.[34]

The carbon dioxide buildup comes from two main sources that both result from the worldwide spread of industrialization and urbanization. The most significant source is fossil fuel combustion, which liberates carbon from coal, petroleum, and natural gas. The other source of carbon dioxide is large-scale deforestation, which transfers carbon from vegetation into the atmosphere through burning and decay. Over the years, much of the increase in carbon dioxide resulted from deforestation, but by the 1980s this accounted for less than 50 percent of the carbon dioxide released.[35]

The impact a doubling of atmospheric carbon dioxide may have on future temperatures is uncertain, since various complex models of atmospheric interactions yield somewhat different results. Generally accepted estimates indicate a global warming of up to 2 degrees centigrade (3.6 degrees Fahrenheit) by the middle of the next century and a major warming of about 5 degrees centigrade (9 degrees Fahrenheit) by the year 2100.[36] Obviously some areas of the planet would be affected by this warming more than others. Changes in temperature will affect transport of water through the hydrosphere and undoubtedly alter rainfall patterns, very likely turning much good agricultural land into deserts and some presently marginal land into lush farmland. Given existing modeling skills, it is difficult to predict precisely agricultural winners and losers, but it is likely that the world will see significant agricultural dislocation and related starvation as long as significant population growth continues to take place at the same time as the global warming.

A more definite—and possibly economically devastating—effect of the projected global warming is a substantial rise in ocean levels around the world. Two factors will be responsible for the rise in sea level. First, as the average global temperature rises, significant quantities of snow and ice now locked in the polar regions will melt and pour into the oceans. Second, the heating of upper layers of water in the oceans will result in a thermal expansion factor that will increase the space the water occupies. It is estimated that warming over the last century has led to an average increase in sea level of between 12 and 15 centimeters.[37] It is much more difficult to estimate future increases because of temperature and other uncertainties, but mid-range estimates indicate that a further increase in sea level of between 8.8 and 13.2 centimeters can be expected by the year 2000 and that an increase of between 52.6 and 78.9 centimeters can be expected by 2050.[38]

The effects of this rise in sea level will undoubtedly be enormous, since small rises in ocean levels can have major impacts on large areas of coastal land around the world. Although an increase of thirty centimeters would hardly submerge the Statue of Liberty, it would erode almost all existing East Coast and Gulf beaches in the United States by more than one hundred feet, and there would be similar impacts in many other low-lying areas of the world. Storm surges would become much more damaging to low-lying coastal communities, and many residents would be forced to relocate. Salt water from the oceans could be expected to move inland, encroaching into estuaries, streams, and groundwater. The economic toll, partially hidden by insurance payments, could be many billions of dollars.[39]

Given that this obvious tragedy is waiting to unfold, it would seem that world leaders would be taking dramatic action to head off damage. But like Hardin's original tragedy of the commons, no single country is willing to take the lead in cutting back on carbon dioxide production. Thus, technocratic exclusionists busily talk up building huge sea walls worldwide to control the rising waters while inclusionists sound like voices in the wilderness stressing the need to eliminate the problem through curbs on carbon dioxide emissions. While learned scientific committees study the dimensions of the problem, the nations of the world continue to add more carbon dioxide to the atmosphere, and the tragedy continues to unfold.

▼ Ozone Depletion

Another human assault on the atmosphere may well be more serious and immediate than the greenhouse effect. While the buildup of carbon dioxide traps infrared radiation in the atmosphere, the ozone layer keeps out damaging ultraviolet solar radiation at the other end of the spectrum. Although ozone makes up less than .01 parts per million of the atmosphere, without it life on the earth's surface would be unlikely to survive. A layer of ozone, beginning about eight miles up in the atmosphere, acts as a protective barrier, filtering out ultraviolet radiation of wavelengths between 290 and 320 nanometers. There is now overwhelming scientific evidence that this ozone layer is under attack from chlorofluorocarbon gases.

Chlorofluorocarbon (CFC) production has risen dramatically since World War II. CFCs are used in refrigeration, air conditioning (freon), as foam-blowing agents, and as aerosol propellants. As production and use of CFCs has risen, more of them have been released into the atmosphere, where they rise into the ozone layer and break down. The breakdown process releases chlorine atoms, which strip away one of the three oxygen atoms making up an ozone molecule. The rate at which this is occurring is subject to a great deal of scientific debate, but even small decreases in atmospheric ozone can have a major impact on human health. It has been estimated that a 16 percent decrease in ozone produces a 44 percent increase in the damaging ultraviolet radiation able to reach the earth's surface.[40] The same 16 percent ozone decrease could result in many thousands of additional deadly skin cancers (melanomas) and significant damage to crops, fish, and wildlife.

No one really knows how much atmospheric loading of CFCs has already taken place, and since CFCs are extremely stable—having lifetimes from about 75 to 100 years—it is difficult to say how close to collapse the ozone layer might be.[41] Scientists have been startled, however, by the appearance and expansion of a large hole in the ozone layer over the antarctic continent each September and October, but they don't understand the mechanism behind this case of ozone thinning.[42]

The potential damage that could be caused by depletion of the ozone layer has spurred U.S. environmentalists to take the lead in suggesting ways to cope with the problem. A major conference was held in Geneva in 1986 in an attempt to get general international agreement on steps to be taken to resolve the issue, and several follow-up conferences have been held. The United States, Canada, and Sweden mandated replacement of CFC aerosols in the late 1970s, but other countries have not been eager to follow suit. The United States has advocated freezing the level of CFC production to be followed by gradual phase-out as substitutes are developed to do the same jobs. But developing and producing substitutes will be expensive; significant numbers of countries balk at the prospect, particularly the less developed countries, which don't want to be bound by production ceilings that would not allow them to increase their production.[43] In September 1987, the United Nations Environmental Program worked out a plan for reducing CFC production thereby protecting the ozone layer, and a treaty to reduce use of CFCs is currently open for signature.

▼ Acid Precipitation

The carbon dioxide buildup is a long-term issue of ecological interdependence; its effects will not be noticed for several more decades. The deterioration of the ozone layer is a much more immediate problem, apparently requiring resolute action within the next decade. But acid precipitation is already doing significant damage and is an irritant in relations among nations. Unlike the other two situations, acid precipitation is not exactly a commons issue because it is not yet global in scope. Rather it is a transborder pollution problem that, for the present, is restricted to limited portions of the atmosphere. It is a major issue in relations between the United States and Canada, and between the Scandinavian countries and continental Europe.

Acid precipitation refers to rainfall, drizzle, snow, and other forms of precipitation that have been made acidic through contact with pollutants, most frequently sulphur dioxide and nitrogen oxides. The greatest share of the former pollutant is created by industry and coal-burning power plants and of the latter by automobile exhaust. Ironically, the long-distance movement of these pollutants has been facilitated by attempts to clean up local pollution by using tall smokestacks at power plants and factories. Once lofted into the atmosphere and carried away from the point of origin by prevailing winds, these pollutants combine with moisture droplets through complex chemical interactions to form acid precipitation.

The acidity of water is measured by its pH value. The lower the pH value, the more acidic the water. On the pH scale an interval of one unit represents a tenfold increase in acidity: thus, 5.0 indicates ten times more acidity than 6.0, and 4.0 indicates one hundred times more acidity. Average rainfall around the world has a value of about 5.0 and that at the South Pole about 5.2.[44] Isolated storms can yield rainfall that is as acidic as vinegar, and certain localities are infamous for high levels of acid in rain. Tampa Bay, Florida, for example, was replaced as an American port of entry for BMW automobiles because of the damage done to exteriors by acid rain. In the most affected, eastern part of the United States rainfall varies between 4.2 and 5.0 in acidity, and in the industrialized portions of Western Europe it varies between 4.1 and 4.9.[45]

Acid precipitation is a hot international issue because there often is a great distance between the source of the pollution and the damage that is done. In the United States it is estimated that one third of the acid precipitation falling in the northeastern part of the country originated at a distance of more than three hundred miles. Thus, the coal-burning power plants of the Ohio Valley are responsible for significant pollution in New England and Canada.[46] In Western Europe, where countries are packed much more closely together, the international exchange of acid pollution is much more serious. In bilateral ecological relations between Germany and France, for example, it is estimated that Germany delivers 124,000 metric tons of sulphur upwind to France each year, while France reciprocates by delivering about 167,000 metric tons downwind to Germany.[47] Acid precipitation moves across borders between the United States and Canada, Western and Eastern Europe, and continental Europe and Scandinavia. Significant activity also takes place in other areas of the world, but monitoring is much less precise.

The United States has been long involved in disputes involving acid precipitation with both its northern and southern neighbors. To the north, Canada has given high priority to the acid rain issue and has placed it at the top of the U.S.–Canadian diplomatic agenda. Canadian officials argue that pollution originating in the United States moves northeastward with prevailing winds and falls as acid precipitation over all of southern Canada. The United States in turn was long embroiled in a diplomatic dispute with Mexico over a copper smelter at Nacozari, just sixty miles south of the border. As originally planned, the smelter would have had no pollution controls, and about five hundred thousand tons of sulfur dioxide annually would have moved north over the border. The issue has apparently been resolved at the highest levels of government in the two countries.[48]

Acid precipitation is a sticky problem for the receiving countries because of its adverse affects on aquatic ecosystems, forests, the built environment, and perhaps on human beings. Over time, acid precipitation can cause the destruction of ecosystems in affected lakes and streams. Damage begins when the pH of water in lakes and streams drops below 5.0. The more acidic water becomes, the more likely the body of water is to be lifeless. Three conditions determine the extent to which a body of water is affected by acid rain: the average acidity of precipitation in a watershed, the types of vegetation and geology through which precipitation passes, and the geological formation in which the body of water is located. Acidic moisture passing through limestone, for example, tends to be neutralized by the alkalinity of the rock and soil.

Evidence of the impact of acid precipitation is incomplete, largely because the extent of the problem is only slowly being recognized politically. In Canada's Province of Ontario it is estimated that more than three hundred lakes have suffered severe damage to fish and other aquatic life.[49] The problem is more severe in Scandinavia, which receives much of the pollution generated over the European continent. Swedish officials estimate that four thousand lakes and ninety thousand kilometers of streams are significantly acidified. The Norwegians claim that thirteen thousand square kilometers of lakes in southern Norway no longer support fish. Fish stocks throughout the region are estimated to be down between 50 and 70 percent.[50] Evidence of the impact of acid precipitation on forests is even more fragmentary, although it is known that trees are damaged both through leaves and needles coming in contact with various pollutants and through changes in the acidity of the soil, which affect root systems. In West Germany recent studies have shown the Black Forest to be turning brown, and 34 percent of all forests there have sustained damage from acid precipitation. More than one billion dollars' worth of trees has been lost because of the phenomenon.[51] There is evidence that Eastern Europe and the Soviet Union have even more severe problems.

Acid precipitation also has an impact on the built environment. It corrodes certain types of building materials and defaces historical monuments. It is suspected that acid precipitation has a direct impact on human health, but evidence on this score is inconclusive.[52]

In the face of this evidence of increasing ecosystem damage, remedial measures would seem urgently required. But because of the transborder nature of the issue—those who profit from the pollution often don't suffer the most adverse effects—there has been no wholesale international attack on the problem. Scandinavian countries have engaged in remedial actions by "liming" affected lakes and have put more stringent limitations on emissions. But since a significant proportion of the pollution comes from other countries, the Scandinavian problem can only be effectively addressed by a multinational effort. In the United States the cost of remedying the situation has been estimated to be between two and eight billion dollars per year.[53] Given the sorry state of the heavy industry located in the so-called rust belt, where much of the pollution is generated, it is unlikely that local industries can absorb the costs of remedial equipment. And given budgetary deficits on the national level, the federal government has been loath to take any responsibility, even though the Canadians continue to raise the issue at every opportunity.

▼ Technology and Other Commons

The positive side of technological innovation has been the industrial revolution and higher living standards for the relevant human populations, but a high environmental price has often been exacted in the course of this industrial progress. In addition to these obvious industrial pollution problems, however, the march of technology has yielded new capabilities creating commons and collective goods problems in other areas. First, technological innovations have improved access to conventional geographic commons such as the seabed and the antarctic. Second, new innovations have opened up the shared electromagnetic communication spectrum to additional uses, and it has become much more heavily burdened over time. Now, its use has become a contentious issue among nations seeking to increase their access to it. Finally, innovations in weaponry have created the potential for a nuclear winter—an ultimate commons tragedy in which the entire global ecosystem could be destroyed by military adventures.

The world's oceans have been an issue among nations for decades, and conflicts over use of these commons culminated in the Law of the Sea Convention of 1982. Historically, arguments about the oceans have focused on rights of innocent passage, the extent of territorial limits, techniques permitted in catching fish, and the fate of certain species of whales. But the rapid run-up of natural resource prices in the 1970s, and the two accompanying oil crises, focused attention on the seabed as a new source of fossil fuels and nonfuel minerals. Technological innovations have made these resources, once thought to be in areas beyond drilling and mining capabilities, potentially available to interested parties. Already, drilling for oil takes place at significant depths off the coasts of many countries. Resource-poor industrial countries covet manganese nodules, small potato-shaped lumps of important minerals, that have been located on the deep seabed. Only the major slump in commodity prices in the mid 1980s dampened potential conflict over access to these resources that have been traditionally considered part of the common heritage of mankind.

The Law of the Sea Convention, which was not ratified by the United States during the Reagan administration, attempts to deal with this commons situation with a unique combination of privatization and collective regulation. Zones of economic jurisdiction are extended out to two hundred miles, effectively eliminating conflict over resources found close to shore. The mining of the seabed has theoretically been put under the control of an international agency that will oversee exploitation activities as well as collect royalties, which are to be used as aid for less developed countries.[54] Obviously the technologically developed countries, such as the United States and Japan, would rather be free to exploit this new resource wealth, and the enthusiasm of these countries for the convention has been weak.

The fate of the antarctic is becoming similarly contentious. For decades the inhospitable antarctic was looked on as an area to be inhabited only by foolhardy explorers willing to endure temperatures that regularly fall below −80 degrees centigrade. But in an increasingly crowded world, the natural resources supposed to be in this region appear attractive. The area south of 60 degrees latitude in the southern hemisphere is a commons protected by the Antarctic Treaty, which entered into force in 1961, when few countries were thinking about potential energy crises and shortages of nonfuel minerals. The treaty governs the activities of the parties in the region for at least thirty years and until such a time as one of the major parties to the treaty requests a review. More than 130 recommendations have also been adopted by the consultative parties to the treaty over time.

The Antarctic Treaty can be expected to come under pressure as the thirty-year period expires in 1991. The twelve consultative parties, who are original members of the treaty, have been joined by four other nations in consultative status. Seven nations maintain claims to portions of the continent, and some of them overlap. And a group of Third World nations would like to resolve the antarctic situation by declaring it to be "the common heritage of mankind," thus putting it beyond national jurisdiction and into the same category as the deep seabed and the moon.[55]

Allocation of the electromagnetic communication spectrum is another technology-induced commons problem. There are only a limited number of frequencies ideally suited for radio and television transmission, and demand for them exceeds supply. In the industrially developed countries, various regulatory commissions have responsibility for parceling out the available slots. But radio transmissions in particular can easily cross national boundaries, thus interfering with radio transmissions in other countries. The International Telecommunications Union, a specialized agency of the United Nations, is charged with overseeing what could be a cacophony of broadcasts by competing stations in different countries. The ITU is a technical body with 158 members, and it has done an admirable job of effectively allocating radio frequencies.

New technologies are presenting challenges to the institutional arrangements for managing this unusual commons. The global telecommunications revolution has given birth to satellite communication, which makes use of geosynchronously orbiting satellites that relay signals from one part of the earth to another. Most of these satellites are located about 22,300 miles above the earth in a narrow band over the equator, where they revolve around the earth in a period of time equal to the earth's rotation. Thus, they appear to remain in one spot and can be depended on to relay signals between points on the earth that can be seen from the satellites. Both the limited number of frequencies used to communicate to and from the satellites and the positions for geosynchronous orbit have been declared to be "limited resources" by treaty.[56]

Growing demands for satellite parking places and related communications frequencies culminated in a World Administrative Radio Conference (WARC) to treat the allocation issue in August 1985. The conference, known as the Space WARC, highlighted deep differences between the industrial and developing countries over allocation policies. The industrial countries, including the Soviet Union, generally support a "first come first served" herdsman-type philosophy whereby the countries with satellite-launching capability could dominate the existing orbit/spectrum resource. Their assumption is that new technological innovations will continually expand this commons in the future, thus assuring less developed countries an eventual opportunity to orbit their own communications satellites. Most of the less developed countries, however, argue for a reservation system that would guarantee them certain existing frequencies and parking places until such time as they might be used. The two differing perspectives have not yet been reconciled, and the Space WARC conference is expected to continue to meet over a period of years.

Finally, the increasing destructive power of weaponry has created a potential ultimate global tragedy of the commons. Advances in nuclear weapons and delivery systems have raised the possibility of a mutually devastating exchange among the major powers that could cause a total economic collapse in the involved countries. Recent research, however, has suggested an even more dire scenario, in which such a major exchange of nuclear weapons would produce side effects that could destroy large segments of the global ecosystem and wipe out the greater part of mankind.[57]

Conventional thinking about nuclear war has focused heavily on the direct impact of explosions on adversaries. Put in an ecological context, however, a significant exchange of nuclear weapons would have global consequences not anticipated by defense planners. A major nuclear exchange, aside from the obvious immediate destruction caused by explosions and radiation, would create four other conditions that would have a long-term destructive impact on the global ecosystem. First, the fireballs from large explosions would create firestorms that could persist for days, if not weeks, since there would be little capability left to fight them. This would release an incredible amount of smoke into the troposphere and would substantially reduce sunlight reaching the earth's surface. Second, the nuclear explosions themselves would hurl dust particles into the stratosphere, in the manner of a major volcanic explosion, and the suspended particles would create a more enduring reduction in sunlight over a large area of the earth. Third, global radioactive fallout would cause various diseases for generations, as wind currents moved the debris to all parts of the world. Finally, thermonuclear weapons produce nitrogen oxides as they explode, and the nitrogen oxides could chemically attack and destroy atmospheric ozone, thus opening up the surface of the earth to a massive bombardment of ultraviolent radiation; this would rapidly accelerate the work that CFCs are already carrying out.[58]

The collection of soot, smoke, dust, and debris lofted into the atmosphere would prevent a significant amount of sunlight from reaching the earth's surface, and a considerable cooling would result. This nuclear winter would not be limited to the countries involved in the nuclear exchange but would be at least hemispheric, if not global, in scope. The intensity and duration of a possible nuclear winter are subject to scientific debate; they would certainly be a function of the scope and length of the nuclear exchange. Some worst-case scenarios project a drop of 50 degrees centigrade in the northern hemisphere for more than six months with only a gradual warming taking place after that. Other estimates, using different climatological models and assumptions, predict a lesser degree of cooling.[59] Obviously these differences cannot be settled by experimentation, although nature has already demonstrated some of the principles involved through major volcanic explosions. The Tambora Volcano in Indonesia exploded in 1815, and the resulting cloud of dust particles in the stratosphere caused a global drop in temperature of 1 degree centigrade. The year 1816 was known as the year without summer in Europe and America.[60]

Regardless of which estimates of the extent of nuclear winter are correct—with any luck scientists will never find out—they mean that a collapse of the global commons would result from a major nuclear exchange. In the northern hemisphere there would be no significant agricultural production at all for at least one growing season, while the impact on production in the southern hemisphere is uncertain. In addition to the immediate utter chaos that would exist in the countries affected directly by a nuclear exchange, there would be medium-term starvation throughout the northern hemisphere and probably the world.[61] And the long-term effects of these changes, and related depletion of the ozone layer, remain to be assessed. Thus, ironically, technological innovations in weaponry have given the world's nuclear nations the potential to act out the ultimate tragedy of the global commons, the integrity of which must be a priority national security concern.

▼ Notes

1. Earl Cook, "The Consumer as Creator: A Criticism of Faith in Limitless Ingenuity," *Energy Exploration and Exploitation* 1, 3 (1982).
2. Glenn Hueckel, "A Historical Approach to Future Economic Growth," *Science* (March 14, 1975); Douglas Bohi and Michael Toman, "Understanding Nonrenewable Resource Supply Behavior," *Science* (February 25, 1983).
3. These arguments were initially laid out in the wake of the first oil crisis. See C. Fred Bergsten, "The Threat from the Third World," *Foreign Policy* (Summer 1973); Zuhayr Mikdashi, "Collusion Could Work," *Foreign Policy* (Spring 1974); Stephen Krasner, "Oil is the Exception," *Foreign Policy* (Spring 1974). For a more recent perspective on this issue see Mark Zacher and Jock Finlayson, *Developing Countries and the Commodity Trading Regime* (New York: Columbia University Press, 1988).
4. See Office of Technology Assessment, *Strategic Materials to Reduce U.S. Import Vulnerability* (Washington, D.C.: U.S. Congress, 1985).
5. John B. Colton, Jr., et al., "Plastic Particles in Surface Waters of the Northwestern Atlantic," *Science* (August 9, 1974).
6. See Sandra Postel, *Water: Rethinking Management in an Age of Scarcity* (Washington, D.C.: Worldwatch Institute, 1984).
7. See Paul Reynolds, *International Commodity Agreements and the Common Fund* (New York: Praeger, 1978), chap. 1.
8. Christopher Brown, *The Political and Social Economy of Commodity Control* (New York: Praeger, 1980), pp. 60–66.
9. See Zuhayr Mikdashi, *The International Politics of Natural Resources* (Ithaca, N.Y.: Cornell University Press, 1976).
10. H. E. Goeller and A. Zucker, "Infinite Resources: The Ultimate Strategy," *Science* (February 3, 1984).
11. Earl Cook, "Limits to Exploitation of Non-Renewable Resources," *Science* (February 20, 1976).
12. Council on International Economic Policy, *Special Report: Critical Imported Minerals* (Washington, D.C.: U.S. Government Printing Office, 1974), p. A-18.
13. Zuhayr Mikdashi, op. cit., pp. 97–102.
14. Earl Cook, "Limits to Exploitation."
15. Council on International Economic Policy, op. cit., p. 13.
16. Raymond Mikesell, *Nonfuel Minerals: Foreign Dependence and National Security* (Ann Arbor: University of Michigan Press, 1987), p. 89.
17. See Charles Park, *Earthbound* (San Francisco: Freeman-Cooper, 1975), pp. 99–103.
18. Council on International Economic Policy, op. cit., pp. A-57–A-59.
19. Christopher Brown, op. cit., pp. 64–66.
20. "Copper's Last-Ditch Plea to Hold Back Imports," *Business Week* (February 13, 1984).

21. Christopher Brown, op. cit., pp. 11–20.
22. See "Tin: How Shearson was Left Holding the Bag," *Business Week* (April 21, 1986).
23. "The Death of Mining: America is Losing One of Its Basic Industries," *Business Week* (December 17, 1984).
24. Office of Technology Assessment, *Strategic Materials: Technologies to Reduce U.S. Import Vulnerability: Summary* (Washington, D.C.: U.S. Congress, 1985), pp. 11–18.
25. Ibid., p. 18.
26. Ibid., p. 16.
27. Mark Schacter, "Cobalt Prices Rise on Concern over Shortages," *The Wall Street Journal* (September 10, 1986).
28. See *Imports of Minerals from South Africa by the United States and the OECD Countries* (Washington, D.C.: Congressional Research Service, 1980).
29. Federal Emergency Management Agency, *Stockpile Report to the Congress* (Washington, D.C.: FEMA, April–September, 1985).
30. Garrett Hardin, "The Tragedy of the Commons," *Science* (December 13, 1968); Garrett Hardin and John Baden, eds., *Managing the Commons* (San Francisco: W. H. Freeman, 1977).
31. "Ground Water Ills: Many Diagnoses, Few Remedies," *Science* (June 20, 1986); Cass Peterson, "Crisis Under Miami: Pollutants Found in Aquifer Holding Drinking Water," *Washington Post* (September 28, 1984).
32. For an analysis of international institutions and environmental issues see Lynton Caldwell, *International Environmental Policy: Emergence and Dimensions* (Durham, N.C.: Duke University Press, 1984).
33. Stephen Seidel and Dale Keyes, *Can We Delay a Greenhouse Warming?* (Washington, D.C.: U.S. Environmental Protection Agency, 1983), chap. 1.
34. A readable explanation of the greenhouse effect is provided by Roger Revelle, "Carbon Dioxide and World Climate," *Scientific American* (August 1982).
35. See George Woodwell et al., "Global Deforestation, Contribution to Atmospheric Carbon Dioxide," *Science* (December 9, 1983).
36. Council on Environmental Quality, *Global Energy Futures and the Carbon Dioxide Problem* (Washington, D.C.: U.S. Government Printing Office, 1981), chap. 1; National Academy of Sciences, *Changing Climate* (Washington, D.C.: National Academy of Sciences Press, 1983); "Trace Gasses Could Double Climate Warming," *Science* (June 24, 1983).
37. John Hoffman et al., *Projecting Future Sea Level Rise* (Washington, D.C.: U.S. Environmental Protection Agency, 1983), p. 32.
38. Ibid., chap. 4.
39. Ibid., chap. 5.
40. National Academy of Sciences, *Protection Against Depletion of Stratospheric Ozone by Chlorofluorocarbons* (Washington, D.C.: National Academy of Sciences Press, 1979).
41. "United States Floats Proposal to Help Prevent Global Ozone Depletion," *Science* (November 21, 1986).
42. "Taking Shots at Ozone Hole Theories," *Science* (November 14, 1986).
43. "United States Floats."
44. *World Resources 1986* (New York: Basic Books, 1986), p. 168.
45. Office of Technology Assessment, *Acid Rain and Transported Air Pollutants: Implications for Public Policy* (Washington, D.C.: U.S. Congress, 1984), p. 66; *The State of the Environment 1985* (Paris: OECD, 1985), p. 33. See also Environmental Resources Limited, *Acid Rain: A Review of the Phenomenon in the EEC and Europe* (New York: UNIPUB, 1983).
46. Office of Technology Assessment, *Acid Rain*, p. 5.
47. *World Resources 1986*, p. 196.

48. "A Mexican Smelter has the Southwest All Fired Up," *Business Week* (July 22, 1985).

49. *World Resources 1986,* pp. 169–170.

50. Ibid., pp. 169–170.

51. Sandra Postel, *Air Pollution, Acid Rain and the Future of Forests* (Washington, D.C.: Worldwatch Institute, 1984), p. 22.

52. "Acid Rain's Effects on People Assessed," *Science* (December 21, 1984).

53. Office of Technology Assessment, *Acid Rain,* p. 19.

54. See Jack Barkenbus, *Deep Seabed Resources: Politics and Technology* (New York: Free Press, 1979); Ann Hollick, *U.S. Foreign Policies and the Law of the Seas* (Princeton, N.J.: Princeton University Press, 1981).

55. Evan Luard, "Who Owns the Antarctic?" *Foreign Affairs* (Summer 1984); Christopher Joyner, "Polar Politics in the 1980s," *International Studies Notes* (Spring 1985).

56. See Milton Smith III, "Space WARC 1985—Legal Issues and Implications" (Montreal: McGill University master's thesis, 1984); International Telecommunications Union, "International Telecommunications Convention," Final Protocol (Nairobi: 1982).

57. See Paul Ehrlich et al., *The Cold and the Dark* (New York: Norton, 1984).

58. Carl Sagan, "Nuclear War and Climate Catastrophe," *Foreign Affairs* (Winter 1983–84), pp. 259–263.

59. Ibid., p. 266; see also Anne Ehrlich, "Nuclear Winter," *Bulletin of the Atomic Scientists* (April 1984).

60. Carl Sagan, op. cit., p. 265.

61. See "After Nuclear War," special issue of *Bioscience* (October 1985).

▼ Suggested Reading

JACK BARKENBUS, *Deep Seabed Resources: Politics and Technology* (New York: Free Press, 1979).

LYNTON CALDWELL, *International Environmental Policy: Emergence and Dimensions* (Durham, N.C.: Duke University Press, 1984).

PAUL EHRLICH ET AL., *The Cold and The Dark* (New York: W. W. Norton, 1984).

GARRETT HARDIN AND JOHN BADEN, eds., *Managing the Commons* (San Francisco: W. H. Freeman, 1977).

JOHN HOFFMAN ET AL., *Projecting Future Sea Level Rise* (Washington, D.C.: U.S. Environmental Protection Agency, 1983).

ANN HOLLICK, *U.S. Foreign Policies and the Law of the Seas* (Princeton: Princeton University Press, 1981).

RAYMOND MIKESELL, *Nonfuel Minerals: Foreign Dependence and National Security* (Ann Arbor: University of Michigan Press, 1987).

IRVING MINTZER, *A Matter of Degrees: The Potential For Limiting the Greenhouse Effect* (Washington, D.C.: World Resources Institute, 1987).

NATIONAL ACADEMY OF SCIENCES, *Changing Climate* (Washington, D.C.: National Academy of Sciences Press, 1983).

NATIONAL ACADEMY OF SCIENCES, *Protection Against Depletion of Stratospheric Ozone by Chlorofluorocarbons* (Washington, D.C.: National Academy of Sciences Press, 1979).

NATIONAL OFFICE OF TECHNOLOGY ASSESSMENT, *Acid Rain and Transported Air Pollutants: Implications for Public Policy* (Washington, D.C.: U.S. Congress, 1984).

OFFICE OF TECHNOLOGY ASSESSMENT, *Strategic Materials to Reduce U.S. Import Vulnerability* (Washington, D.C.: U.S. Congress, 1985).

SANDRA POSTEL, *Air Pollution, Acid Rain and the Future of the Forests* (Washington: Worldwatch Institute, 1984).

STEPHEN SEIDEL AND DALE KEYES, *Can We Delay A Greenhouse Warming?* (Washington, D.C.: U.S. Environmental Protection Agency, 1983).

RAE WESTON, *Strategic Materials: A World Survey* (Totowa, N.J.: Rowman and Allenheld, 1984).

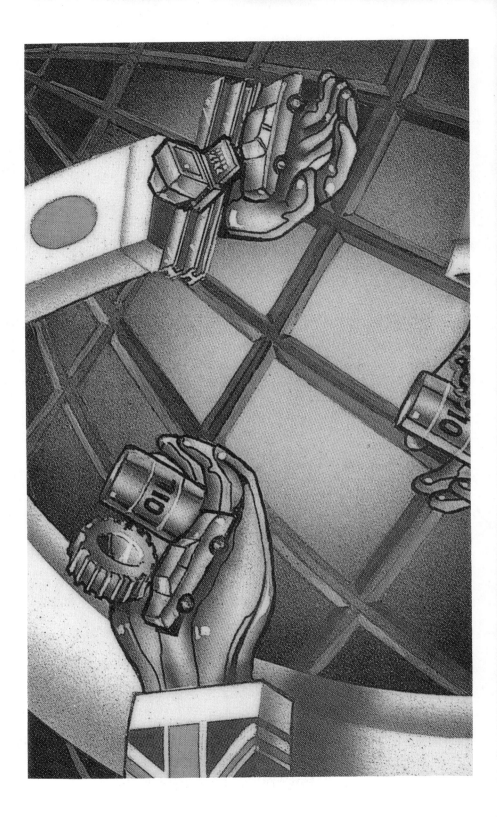

6

Technology
and the
Structure of
Global Inequality

The uneven spread of the technologies of the industrial revolution has created a world that is sharply divided between rich and poor. Material-intensive industrial technologies first burst forth in Western Europe and then spread slowly outward to other countries. The pace and manner in which this incomplete revolution spread have created numerous technology-related problems between the countries that developed early and those still in the beginning stages of the industrialization process. Development of new technologies permitted the early industrializers to establish colonial empires to meet the growing material and labor needs of the industrial revolution. The persistence of technological differences now stacks the deck against these former colonies in their developmental efforts. Technology creates strong competitive advantages for the early industrializers in the global economy; the late developers largely remain, as they were in colonial times, exporters of minerals and agricultural commodities.

Theorists writing from a Third World perspective attribute most of the problems the contemporary less developed countries face to capitalism and a capitalist world system. Evil men and their capitalist institutions established the dominance of core countries over the periphery, and through their machinations now maintain a neocolonial hold over these countries.[1] This analysis of many of the problems faced by these countries may be accurate, but it still misplaces the blame for developmental dilemmas. Technological differences established the colonial empires of the industrial period, and these same differences provide the underpinnings of contemporary neocolonialism. In other words, from an inclusionist perspective, the cause of contemporary developmental differences lies not in evil men with malevolent values but in deep differences in technological sophistication, which stack the deck against the late industrializers.

The most commonly accepted liberal economic prescription for closing the yawning gap between the early and late industrializers is unrestricted international trade. After decades of efforts to follow these prescriptions, the development record is at best mixed. Free-market economists continue to tout the virtues of removing all trade barriers in less developed countries and point to the successes enjoyed by a handful of newly industrializing countries (NICs) in Asia. But on closer examination, free-trade philosophies tend to benefit technologically superior countries, except for a few special cases where developmental timing may have facilitated rapid industrialization. The free-trade push of the last two decades has resulted in very little per capita progress for many of the world's less developed countries, and few of them give indications of emulating the NICs. Development economists are thus now going through major reappraisals of development strategies, and prescriptions for economic growth are currently in flux.[2] Because of the persistent nature of the problems associated with the development gap and the increasing numbers of human beings living in the less industrialized world, the development dilemma is likely to grow in importance well into the next century.

The question of winners and losers in the international development game is a complex one, but it is clear that the onward march of technology is creating an underclass of nations with poor competitive prospects, which is often referred to as the never-to-be-developed world (at least never to be developed on the basis of old international economic order prescriptions). A handful of countries clearly has the wherewithal to make it by following the liberal development path, but reservations in the club are now limited to a select few.

Those adhering to shopworn development models have assumed that the gap between rich and poor would narrow as a natural outcome of worldwide industrialization. Considerable evidence has mounted, however, showing that the absolute size of the gap between the richest and poorest countries is widening instead. In the quarter century following 1960, real GNP per capita in the OECD countries nearly doubled, while that in the less developed countries increased by only 86 percent. The result is that real OECD incomes per capita increased from $5304 to $10,220, while the less developed countries were able to move only from $391 to $729 during the same period. OECD countries grew at an annual rate of 1.1 percent between 1980 and 1984, while the developing countries didn't grow at all. In fact, real per capita GNP dropped 2.5 percent per year in Africa and 2.6 percent per year in Latin America over the period in question and has been pretty much stagnant since.[3]

From an inclusionist perspective, inequities in technological innovation, population growth, and resource distribution can be seen as significant factors in the origin and persistence of the development gap. The fossil fuel–intensive technologies of the industrial revolution originated in Western Europe and only gradually and unevenly spread outward to various other parts of the world. The early industrializers used the power of these new technologies to forge colonial empires, a response to the lateral pressure of the times. But technological innovation has remained concentrated in the countries that industrialized early, thus making labor and resource utilization much more efficient there. The old economic growth model assumed a rapid diffusion of technological innovation to the less developed countries that would permit those countries, using cheap labor and resources, to become active and successful participants in the global economy. Except for a few fortunate countries that have managed to find unique development niches, however, successes have been modest, and in many cases there has been a pronounced lack of progress. Technological innovation in many of the already industrialized countries has accelerated, and automation has at least partly offset the cheap labor edge enjoyed by the less developed countries.

Rapid population growth in the LDCs, treated in detail in Chapter 2, has also been responsible for the widening gap. Modernization is a process consisting of many components, and it hasn't always progressed evenly. The early stages of change are dominated by the introduction of new technologies, many of which directly or indirectly lengthen the life span, decrease infant mortality, and lessen the scourge of disease. The structural and value changes that accompany the industrial modernization process are less readily accepted and lag generations behind. As a result, the pronatalist values of an agrarian society often persist in a world in which more children survive and people live longer—the origin of the contemporary population explosion. In many of these countries populations are doubling every eighteen to twenty years, and strenuous economic growth efforts are required simply to keep from losing ground. The early industrializers never experienced such rapid population growth because the modernization process was indigenous and took place over many genera-tions. In contemporary LDCs, the impetus for change comes from abroad, and given the existing revolution of rising expectations, there is inadequate time for the process to unfold in an orderly fashion.

Natural resources provide both development opportunities and constraints in the less developed countries. Most of them are net importers of petroleum and were negatively impacted by the twin oil crises. While many LDCs are exporters of raw materials, they compete with one another in marketing undifferentiated basic com-modities, and their earning power is quite limited, especially during periods of slow global economic growth. Producers of bauxite and copper and growers of coffee, sugar, and cotton, these countries have suffered from serious commodity price in-stability as well as long-term deterioration of trade relationships. And technological innovation in the already industrialized world threatens to limit future markets for their basic commodities by means of high-technology substitutes. Even more impor-tant, the raw material demands of economic growth have seemingly peaked in the United States, Western Europe, and Japan as a result of the post-industrial transfor-mation of economic activity now under way. Material substitutions. design changes, saturated markets, and a new mix of products are all combining to limit future demands for exported raw materials.[4]

▼ Technology and Inequality

The countries that industrialized early—Great Britain, Germany, the United States, and eventually several other OECD countries—continue to maintain significant technological advantages over countries that are developmental latecomers. Industrial technologies, feeding on cheap resources, created social surplus of unprecedented magnitude in the countries that first industrialized. This surplus both fostered fur-ther rounds of innovation and permitted the development of global empires. These core, early industrializing states were able to use their peripheral colonies as cheap sources of natural resources and labor, thus accelerating their own industrial growth. In most cases the colonizers did very little to prepare colonies for eventual political independence or economic competition. Now that these peripheral states have at-tained independent status, the remnants of the colonial era remain as obstacles to economic growth. In the existing economic order the technologically sophisticated countries possess a tremendous competitive edge across the board, and this edge is important in structuring global inequality.

Table 6-1 highlights some of the technological factors underlying the poverty gap. Competitiveness in the contemporary world marketplace requires investment in education and in research and development. The evidence indicates that education and scientific and technological expertise are heavily concentrated in the core coun-tries that industrialized early. The developed world, for example, averages 2875 scien-tists and engineers for each million people, while the less developed world has only 121. There are also huge gaps in spending on research and development. Although it

Table 6-1 The technology gap. Column 1 is per million, 1980; column 2 is billions of dollars, 1980; column 3 is per inhabitant, 1985 (public education).

	R and D scientists/engineers	R and D expenditures	Education expenditures
World	848	208	147
Developed	2986	195	515
Developing	127	13	27
North America	2679	67	1101
Europe	1735	71	294
Africa*	91	1	37
Asia	271	31	46
USSR	5172	32	**
Latin America	251	4	67
Oceania	1472	2	436

*Includes Arab African states.
**Included with Europe.
SOURCE: *UNESCO Statistical Yearbook 1987.*

is difficult to gather such data, UNESCO estimates that in recent years the world has spent an annual average of about $208 billion on research and development. The developed world accounts for about 94 percent of this spending and the less developed world for only 6 percent. Similarly, the developed world spends about $465 per inhabitant on education, while the less developed countries manage only $27. About 90 percent of the world's research and development scientists and engineers work in the industrial countries and only 10 percent in the LDCs.

The countries on the other side of the technology gap are often collectively referred to as the Third World. As the data in Table 6-2 indicate, the Third World is really a very complex mix of countries ranging from the newly industrializing countries to the economic basket cases charitably called the low income economies. In an effort to make order out of this chaos, the World Bank divides the Third World into four groups of countries ranging from the wealthy, but poorly developed, high income oil exporters to the very poor, low income economies. Evidence of the persistence of global inequality is most vividly seen in comparisons between the nineteen industrial market economies and the thirty-four low income economies. China and India, with their massive populations and low income levels, would technically be in the latter category, but they are generally separated out for special consideration because their performance differs markedly from the other thirty-four, many of which are found in the poverty belt in sub-Saharan Africa.

The figures indicate a significant and growing gap between the very rich and the very poor countries. The average income of a person in the industrialized countries is well in excess of $11,000, while people in the thirty-four low income economies average a meager $190. Over the last twenty years the industrial economies have grown at a real per capita rate of 2.4 percent each year, while the least industrial economies have been plodding along with growth of less than 1 percent. These figures translate

Table 6-2 The development gap.

	1985 GNP per capita (U.S. $)	1965–85 real GNP growth per capita (%)	1985 infant mortality (per 1000 births)	1985 life expectancy (years)	1985 daily calorie supply per capita
Industrial market countries (19)	11,810	2.4	9	76	3417
Upper middle income countries (20)	1850	3.3	52	66	2987
Lower middle income countries (40)	820	2.6	82	58	2514
Low income countries (35)	200	0.4	112	52	2073
China and India	290	3.5	58	63	2428

SOURCE: World Bank, *World Development Report 1987.*

into considerable differences in the experienced quality of life. Children born into the industrial market economies can expect, on the average, to live for 76 years, whereas their low income counterparts can expect only 52 years. People in the industrial countries eat very well, consuming an average of 3352 calories of food energy per day, about 130 percent of minimum needs. On the other side of the gap, the average calorie consumption is 2275, almost exactly the minimum daily requirement. In many of these countries, of course, the average calorie consumption is much lower, averaging only 90 percent of basic needs in sub-Saharan Africa.[5]

Conditions are so chaotic in the low income countries that in seven of them adequate economic data are not available. In six of these low income countries the economy has actually been losing ground to population growth over the last two decades, and in another six growth per capita has been less than 1 percent per year. In only seven of these countries is the daily calorie supply per person above the minimum. Although life expectancy at birth now averages 52 years, a person born in Guinea can expect to live only 38 years.

There are many reasons that the low income and lower middle income countries did not fulfill hopes for their economic progress during the two UN development decades that terminated in 1980. Most of these reasons are related to the historically slow spread of industrial technology to them. Very little expertise was transferred from the core to peripheral areas during the colonial period. On attaining independence the peripheral countries continued to provide natural resources to the industrial world, since they lacked production facilities themselves. Lack of technological sophistication still keeps these countries from actively competing in an ever-changing export market. To make matters worse, foreign aid has increased at a rate not nearly adequate to keep up with Third World needs. And finally, past borrowing from the private sector combined with rapidly fluctuating interest rates and currency values has created repayment problems that act as a further brake on economic growth. Finding ways for these countries to break out of the poverty syndrome and avoid economic instability and political revolution will be one of the primary challenges of the twenty-first century.

▼ The Colonial Legacy

During the wave of technology-driven outward expansion that accompanied the industrial revolution, only modest concern was evinced for the future welfare of the inhabitants of the colonies that were eventually to become independent nations. They were regarded as extensions of the sovereign territory of the colonizing power and as sources of raw materials and labor for meeting the needs of growing economies. In most cases the colonial powers were loath to locate any downstream production facilities in the colonies or otherwise transfer technological expertise to these agrarian areas of the world. Thus, the industrial revolution that spurred growth in Western Europe and the United States never really took hold in these captive areas of the world; on attaining independence, these new nations possessed very little technological expertise or social and economic infrastructure to launch their own industrialization.[6]

The legacy of the technological differences that structured relations during the colonial period remains a factor in current relations between the industrial core countries and their less fortunate peripheral counterparts. Most of the colonies were developed as one-crop economies, producing only one or two main crops such as coffee, tea, cocoa, or sugar. Other colonies were exploited as sources of valuable metals, such as silver or gold, and eventually less precious metals. Since most raw materials could be transported to and processed by factories in the colonizing countries, there was little need to train the colonized labor force for anything but menial labor.

The failure of the colonial powers to transfer technology and expertise and to locate significant downstream facilities in most of their colonies has left a legacy of fundamental competition problems for the LDCs. Complex and expensive industrial production facilities, continued technological innovation, and the economies of scale characteristic of large industrial countries give the early industrializers a seemingly insurmountable edge in efficiently transforming natural resources into products for world markets. The less developed countries, in general, must compete without having adequate facilities, enough expertise, or large domestic markets to enable them to gain an economic toehold in international competition. Thus in many cases they remain purveyors of labor-intensive products and primary commodities.

As the industrial revolution unfolded, it created growing demands for raw materials in the industrializing countries, and a steady flow of resources from less to more industrialized countries developed. These flow patterns, characteristic of the colonial era, have persisted to the present and are often referred to as constituting neocolonialism (see Figure 6-1). The low income economies generally export raw materials, and the economically developed countries export high value-added products. Technology gives the industrial countries a comparative advantage in almost any economic sector they choose to exploit, ranging from substitution of fiber-optic cable for copper wiring to development of artificial sweeteners to replace natural sugar.

Now that the peripheral countries have become politically independent, they are locked in heated competition with one another to expand exports of primary commodities into limited world markets. Particularly in light of very weak global economic growth in the 1980s, these markets are now saturated with raw materials. Furthermore,

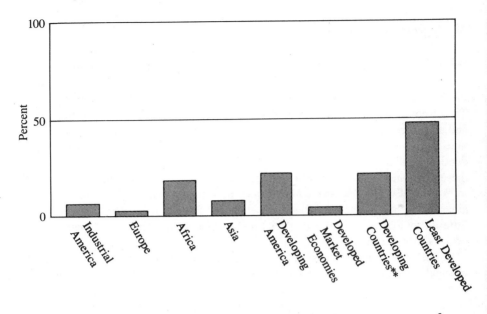

Figure 6-1 Basic commodity exports: eighteen basic commodities as a percentage of merchandise exports, 1983.
*Excludes major petroleum exporters.
SOURCE: United Nations, *Yearbook of International Commodity Statistics 1985.*

as industrial customers for primary commodities move into a less material-intensive post-industrial period, demand for raw materials drops sharply.

Not only must low income countries squeeze meager export earnings out of agricultural commodities, but high-technology farmers in industrial countries, often subsidized by their governments, have made significant inroads into even these relatively stagnant markets. For example, more than one fifth of all exported rice has come from the United States in recent years, a situation that developed largely because of subsidized production in the United States, where mechanized operations, sometimes including helicopters, are frequently used. A similar situation exists in sugar production: the U.S. market has been carefully protected by quotas, and U.S. farmers are paid more than three times the world market price to grow sugar beets.

A number of low income countries export nonfuel minerals, but they are bedeviled by problems similar to those that trouble exporters of agricultural commodities. For example, Zambia, Chile, Zaire, and Peru dominate the world copper market, sharing nearly one half of all exports among them. But copper also comprises the largest share of their export earnings, which makes them very vulnerable to price fluctuations. Other less developed countries export bauxite, iron, and a host of other minerals and are often equally dependent on one or two minerals for most of their export revenue. Finally, almost all exported ore is marketed through multinational corporations, which control downstream facilities and thus are able to control prices.[7]

Low income countries have attempted to remedy some of these structural handicaps politically by forming producer associations and demanding that an Integrated Program for Commodities (IPC), which would theoretically stabilize prices for basic commodities, be established through the United Nations. Some progress has been made in product diversification, but the commodities slump of the 1980s severely hampered producer associations and related commodity agreements with importing countries.[8] The long-standing International Tin Agreement collapsed because of a market glut in 1986, and the international coffee market was in chaos because of over-production in 1987. While compensatory schemes such as the Lomé Convention, a commodities agreement between Common Market countries and former colonies, have been put in place on a regional basis for some commodities, there has been little support from the United States or other industrial countries for a broader Integrated Program for Commodities. In fact, there are major problems with such compensatory schemes because without tightly enforced quotas, higher prices lead to overproduction.[9]

Both the business cycle in industrial countries and global weather fluctuations create market uncertainties and rapid price fluctuations that are damaging to one-crop economies. Agricultural commodity prices are often unstable because they are driven by changing weather conditions. A frost-damaged coffee crop in Brazil, for example, can cause coffee prices to skyrocket on world markets, much to the benefit of other coffee exporters. But an outbreak of good weather in major exporting countries can cause gluts and price collapses. Similarly, mineral exporters are periodically victimized by economic slowdowns in the industrial countries. A burst of economic activity in industrial countries causes markets to tighten and prices to rise. But economic slowdowns depress raw material demand and cause commodity prices to tumble. Moreover, mineral exporters are continually adding new mining capacity in their attempts to gain a greater market share, thus creating excess supply and depressed prices.

A long list of additional handicaps inherent in the colonial heritage frustrates efforts by former colonies to close the development gap. Railroads and highways in these countries were designed and built by the colonial powers with an eye to facilitating the flow of raw materials from inland areas to ports. Thus, the old transportation networks that lead to the sea had to be augmented by expensive new infrastructures to better meet domestic developmental needs. Even the national borders of many of these countries, imposed by the colonial powers, make little political or economic sense. Often the same political units encompass antagonistic tribal or ethnic groups, because boundaries were drawn where colonial spheres of influence intersected. The colonial powers gave little thought to the long-term political consequences that would ensue from diverse tribal, ethnic, or linguistic groups being trapped within the confines of a single state. Thus when Nigeria became independent of Britain in 1960, the country fell almost immediately into a civil war. By 1966 the Ibo tribe became emboldened to attempt a coup—an attempt quickly thwarted by the violent reaction of the Hausa, Fulani, and Yoruba tribes, which massacred tens of thousands of Ibos. Tribal conflict, reinforced by religious fundamentalism, remains a potent and destructive force in contemporary Nigeria. These types of internecine quarrels remain common in many less developed countries and are a major barrier to the political stability needed for economic development.[10]

▼ The Changing Environment for Growth

Another set of factors perpetuating the development gap is related to the increase in interdependence that has taken place since the early industrializers began the growth process. The increase in various forms of interdependence and changes in the rules of the development process have created a very different environment for contemporary development efforts, but political leaders in the countries that industrialized early refuse to acknowledge or compensate for these differences.

The most obvious change in ecological interdependence is in the relationship between global industrial growth and the deterioration of the physical environment. A global ecosystem that will soon sustain five billion human beings is under much greater environmental stress than the one that sustained only five hundred million. Political leaders in the countries that industrialized early were able to respond to growing lateral pressure by moving across borders to establish jurisdiction over thinly populated areas of the world. Today, however, leaders of demographically exploding less developed countries have no such possibilities. There is little productive empty land left in the world, and there are few lightly defended borders across which to move.

Increasing natural resource costs and related limits are other ecological factors reducing growth potential in many of the low income economies. The early industrializers had access to cheap supplies of fossil fuels, primarily coal, and could use the natural subsidy these fuels offered to spur economic growth. In the contemporary less developed world, many countries lack the fuels and minerals essential to economic growth. And better understanding of the costs of environmental pollution make coal a questionable fuel to use wherever it is available. Claims made by some LDC leaders that their land has been raped and pillaged during the colonial occupation ring hollow; but it is true that the low income countries have not been generously endowed with cheap and clean fossil fuels, and the run-up in oil prices on the world market during the OPEC decade dealt a serious blow to development aspirations. Even such primitive sources of energy as firewood are scarce in many countries and are disappearing in the face of rapidly growing populations.[11]

Finally, growing pressures on nonconventional resources in the global commons have created new environmental awareness and an emerging set of environmental values and standards. Early industrialization took place with little concern for environmental integrity. For instance, the classic fogs of industrializing London were actually a pall of smog cast over the city by unregulated combustion of wood and coal. Today, however, pollution is seen to be a global problem, and remedies are to be found by expensive treatment of the pollution that was previously taken to be an externality. Although environmental conditions in the less developed countries may seem appalling by newly adopted standards, they are probably not very different from those that accompanied early industrialization. But there are now growing external pressures on the less developed countries for costly environmental cleanups that, though certainly needed to avoid serious environmental harm, are an added obstacle to rapid industrial growth.[12]

The growth of structural interdependence has also worked to the detriment of the late developers. The early industrializers experienced relatively slow and autonomous growth, free of challenges from foreign competitors. Since there were no external models of affluence to induce relative deprivation, plenty of time was available to work out the dramas and crises of modernization. With no imperatives and pressures to catch up with other countries, substantial increases in gross national product per capita could take place across centuries rather than decades. But today, the interdependent world is filled with competitive hazards for the late industrializers. International markets are glutted with products from more technologically sophisticated countries, making it difficult for the late bloomers to break into them. As a result of a special set of circumstances, a handful of countries, the newly industrializing countries, have become competitive. But for the most part, even cheap labor cannot help the low income countries move into more lucrative export markets.

Growing value interdependence can also adversely affect growth processes in low income countries. Capital accumulation in the countries that developed early was accompanied through low wages, child labor, sixty-hour work weeks, primitive working conditions, and factory dormitories. Standards have changed over time in response to growing affluence. Now, value chauvinists in industrial countries seek to impose their new standards on the contemporary less developed countries. Thus, human rights advocates in industrial countries regularly berate the LDCs for low wage rates, long work weeks, and child labor. Most people would like to see a world where everyone received a high minimum wage and worked a forty-hour week, but to implement these kinds of universal standards would take away the one set of production advantages that is enjoyed by the less developed labor-intensive economies.

▼ Accumulating Capital

To begin to close the gap between their nations and the industrial countries, leaders in the low income countries must find ways to encourage people to forgo current consumption in the interest of capital accumulation for future economic growth. Under current adverse conditions, however, this is unlikely, if not impossible. Rapid population growth and related poverty in the LDCs make it very difficult to get people to save for the future. Although economic growth can be accelerated by importing expertise and equipment from abroad, this developmental strategy normally requires hard currency. Leaders in the low income countries thus not only need to find formulas to squeeze capital for future growth domestically, they also must acquire capital from abroad to pay for needed imports. Both tasks are associated with many potential political pitfalls.

Numerous recipes for obtaining capital have been written by development economists over the years, and politicians employing them have had very mixed successes. Ul Haq listed several phases through which development thinking has passed in attempts to find such a magic formula. Initially, economic planners stressed import substitution as a way of husbanding scarce foreign exchange. But then fashions changed, and less developed countries were urged to expand exports. In more recent years, leaders of less developed countries have been told to pay more attention to domestic agriculture, to control population growth, to open up markets to free enterprise, and to redistribute wealth.[13] Each of these strategies has worked for some countries for a short period of time, but in most cases growth has eventually slowed, and the development gap has continued to widen.

Basically, poorer nations can follow four avenues in attempting to obtain capital to sustain industrial growth. First, given large domestic markets and time, they can attempt to squeeze some surplus out of a hard-pressed domestic work force. This self-reliant method of development might work for some large countries that have the luxury of time. In the contemporary world, however, very few countries are large enough to profit from such strategies, and almost no political leaders have the luxury of time. Both Tanzania and China tried self-reliance for considerable periods over the last two decades; the results were not particularly impressive. Because of dismal results, Tanzania has been backing steadily away from self-reliant development. The Chinese, with a bigger domestic market and associated economies of scale, have had more success, but the opening to the outside world in the 1980s substantially improved prospects for industrial growth.

The other ways capital can be obtained for development involve enhanced proceeds from trade, governmental foreign aid, and development loans. Most less developed countries have attempted to tap all three sources with very mixed results. Some—for instance, the newly industrializing countries in Asia—have done very well with export promotion and seem to be on the way to prosperity. But others have not been as fortunate, and over the last decade capital from all three sources has become much more difficult for many LDCs to obtain.

The remarkable success of the Asian NICs—the so-called Gang of Four (Korea, Taiwan, Hong Kong, and Singapore)—has fostered faith in export promotion as a prescription for trading to prosperity. Careful analysis of the NIC experience, however, raises questions about whether their successes can be replicated within the tighter constraints of the contemporary world economy.[14] An unusual symbiotic interaction between NIC export-led growth and rapid expansion of world markets made this form of industrial development work during the late 1950s and early 1960s.[15] But in an anticipated future world market characterized by limited growth and a surge in production capacity, the rules for success will have changed dramatically.[16]

A number of factors already limit possibilities for rapid export-led growth. The most important is that low income countries, at least partly as a result of the colonial legacy, mainly compete with one another to export raw materials. At least forty-eight less developed countries receive more than three quarters of their export revenue from primary commodities, and most receive over half.[17] This creates two types of problems for countries on the low end of the development gap. First, most basic commodity prices have deteriorated relative to the prices of manufactured goods over time, thus adversely impacting trade receipts in less developed countries. Second and perhaps most important, prices of these commodities have fluctuated wildly over the last two decades, making long-range developmental planning difficult.

The economic slump of the mid 1980s did severe damage to plans for growth through export promotion. Lagging demand for tin, combined with an excess of mining capacity, caused the long-standing International Tin Agreement to fall apart and sent tin prices into a tailspin in 1986. The International Coffee Agreement also came unraveled in a quarrel over the allocation of quotas in a glutted market in 1987. Thus, the wholesale price of coffee, which peaked at $2.29 a pound in 1977, dropped to $1.20 a pound a decade later. Other commodities of importance to the economies of less developed nations have also done poorly. Sugar sold for forty-four cents a pound on the world market in 1980 and dropped to five cents a pound by 1985, and copper hit $1.01 per pound in the former year before dropping to sixty-six cents in the latter.[18]

Technological innovation in industrial countries also diminishes export potential for basic commodities. Sugar is one of the most important exports from low income countries, but the market for it has shrunk significantly subsequent to the introduction of low-calorie artificial sweeteners. Rubber bushes have been developed to grow in desert regions of the United States, possibly opening up competition with natural rubber produced in LDCs. New breakthroughs in biotechnology may have even more serious consequences for LDCs. The juice-producing vesicles of oranges, lemons, and citrons, for example, have been separated out and grown to maturity under laboratory conditions, thus potentially obviating the need for trees. Some day citrus factories in industrial countries will produce fruit juices, ridding those countries of the need to import citrus fruit from LDCs.[19] And last, as the technologically advanced countries continue to benefit from post-industrial or third-wave technologies, their demands for raw materials will level off or even decline. Overall demand for raw materials will diminish because of declining production of goods in service societies and because new materials, such as fiber-optic cables, can replace the traditional minerals exported by less developed countries.[20]

Finally, even politics in the industrial countries works against the interests of export-oriented LDCs. In both the United States and the Common Market, farmers are paid handsome subsidies to grow various crops, including the sugar that is such an important LDC export. In the mid 1980s, the world market price for sugar was five cents per pound, but European farmers were paid eighteen cents per pound and American farmers twenty cents to grow for the domestic market. Furthermore, both the Common Market and the United States protect domestic markets with quotas. In 1986, for example, the United States gave farmers a boost by reducing the imported sugar quota from 2.4 to 1.7 million tons. More important, in recent years additional political pressures for protection from cheap imports in the United States have closed off potential markets for semifinished products, thus penalizing less developed countries making some strides toward product diversification and attempting to get hard currency to service their debts to American banks. And last, in a situation filled with pathos and irony, U.S. antidrug policies, forced on a number of less-than-willing western hemisphere countries, have a devastating economic impact on the exporting countries. Although no one would advocate development through drug sales, it is a fact that countries like Jamaica, Colombia, and Bolivia are dependent on marijuana and cocaine production for as much as one half of total gross national product and export earnings. Crop eradication efforts are necessary and laudable activities but nevertheless have a depressing impact on the economies involved.[21]

Foreign aid—official development assistance—is another channel through which low income economies can get access to growth capital. But there are a lot of misconceptions about the usefulness of aid, the amount given, and the reasons it is given. Donor countries supply assistance for many reasons, but an analysis of recent patterns of aid shows humanitarian concerns to be toward the bottom of the list and political and strategic considerations at the top. During the first UN development decade, the 1960s, it was suggested that OECD countries contribute at least .7 percent of gross national product in development assistance. Very few countries ever reached that target (see Table 6-3). The percentage of GNP given by the United States in particular has been declining over the last two decades. Not only have many countries been lax in the amount contributed, but political factors seem to play a large role in selection of recipients for the contributions they do make. For example, in the early 1980s nearly 40 percent of bilateral U.S. aid went to Egypt and Israel, 38 percent of French aid went to four overseas departments and territories, and 42 percent of all OPEC aid was sent to Jordan and Syria.[22]

The questions whether aid should be given for humanitarian purposes and whether governments are motivated by these considerations are very complex.[23] The

Table 6-3 Official development assistance.

	Millions of U.S. $ (1986)	% of Donor GNP			
		1965	1975	1980	1986
Norway	797	.16	.66	.85	1.20
Netherlands	1738	.36	.75	1.03	1.00
Denmark	695	.13	.58	.74	.89
Sweden	1128	.19	.82	.79	.88
France	5136	.76	.62	.64	.72
Australia	787	.53	.65	.48	.49
Belgium	542	.60	.59	.50	.48
Canada	1700	.19	.54	.43	.48
Finland	313	.02	.18	.22	.45
Germany	3879	.40	.40	.44	.43
Italy	2423	.10	.11	.17	.40
United Kingdom	1796	.47	.39	.35	.33
Switzerland	429	.09	.19	.24	.30
Japan	5588	.27	.23	.32	.28
New Zealand	66	—	.52	.33	.27
United States	9784	.58	.27	.27	.23
Austria	197	.11	.21	.23	.21

SOURCE: World Bank, *World Development Report 1987.*

total amount of aid from OECD countries has been pretty much on a plateau since 1980, when \$27.2 billion was transferred to middle and low income countries. The OPEC countries gave considerable aid when oil prices were high, but contributions have dropped significantly along with prices. Measured OPEC assistance peaked at \$9.7 billion in 1980 and fell dramatically after that. The United States is credited with being the biggest gross aid donor over recent years, but this is to be expected from a country that has by far the world's largest economy. In 1965 the United States was giving .58 percent of its gross national product in foreign aid, but this figure plummeted over the years to .23 percent of GNP, the smallest relative contribution of any OECD country.

Aid data illustrate the political and strategic factors that determine who receives it. Egypt and Israel, potential adversaries as well as strategically important allies of the United States, receive about one eighth of all world foreign aid given. India and Pakistan, potential enemies, also receive a great deal of aid. Although both countries could make a legitimate claim for humanitarian assistance, possible conflict between them as well as their strategic locations are certainly factors that have influenced donors. Jordan and Syria are also high on the recipient list for strategic and political reasons, both being heavily supported by OPEC countries. Israel and Jordan are hardly the two most needy countries in the world, making their per capita support received from aid quite astonishing. Thirty-four low income countries, by contrast, received only \$8.7 billion in aid in 1984, which was down from \$9.2 billion in 1980. Net bilateral flows of aid to these countries equaled only .07 percent of OECD GNP in 1984, way down from .20 percent in 1965. The United States only contributed a net .03 percent of GNP to these countries, hardly a step toward closing the growing poverty gap.[24]

Recent U.S. foreign assistance has been driven by military and political considerations. In 1983, for example, the United States supplied 96 percent of all aid received by Israel, 68 percent of all aid received by Egypt, and 51 percent of all aid received by Turkey.[25] U.S. aid is also more heavily focused on guns than on butter. Only about one third of U.S. aid given between 1978 and 1986 was development and humanitarian aid, while the rest, including the economic support fund for countries with instability problems, was basically security assistance.[26] (See Table 6-4).

Table 6-4 Major recipients of U.S. loans and grants 1946-1986 (net aid in billions of dollars). (Vietnam also received significant aid.)

Egypt	13.1
Israel	12.7
India	11.4
Pakistan	6.6
South Korea	6.1
Turkey	4.2
Indonesia	3.4
Philippines	3.1

SOURCE: Central Intelligence Agency, *Handbook of Economic Statistics 1987.*

In summary, relatively little aid seems to be given to the least developed countries for humanitarian purposes. Much of the aid that is given, particularly by the United States, is allocated to client states for strategic purposes. Furthermore, while populations have been rapidly growing in the low income countries, foreign aid measured in constant dollars has been plunging. Thus, aid as a source of capital accumulation is much overrated. In an era of limited opportunities, many comfortable assumptions about the potential of aid for closing the gap between rich and poor countries have been shaken by the unwillingness of developed countries to make sincere aid commitments.

▼ The Debt Trap

Given the political and economic limitations of foreign aid in relation to the growing demands for it, private sector financing has emerged over the last decade as an alternative. The private sector supplies development capital in two different ways. Foreign direct investment, which involves building facilities and locating operations in less developed countries, has been the traditional method of transferring capital, technology, and expertise. More recently, however, commercial banks have become more actively involved in direct lending to governments and corporations in less developed countries and have filled some of the lending void.

Multinational corporations, seeking opportunities for high rates of return on investment, have historically invested in carefully selected less developed countries. Cheap labor and abundant resources have been important factors motivating corporations to locate significant production abroad. The extent of multinational investment by U.S. firms was so large in the late 1960s and early 1970s that they were heavily criticized for influencing politics and economics in host countries.[27] In the 1980s, however, the investment tide turned considerably, and leaders of many Third World countries offered tax breaks and other inducements to attract direct foreign investment. In the period 1960–65, 20 percent of all capital flow to less developed countries was in the form of direct foreign investment. This dropped to 13 percent for the period 1980–85.[28] The lure of cheap labor and resources has proved to be short-lived for multinational corporations in the face of chronic political instability, nationalization of facilities, poorly educated labor, currency controls, and trade barriers in many potential host countries.

The big increase in development funding has come in the form of commercial lending. Official development assistance made up 50 percent of LDC capital flows and 78 percent of capital flows to low income countries in 1970. By 1983 this had dropped to 46 and 45 percent respectively. At the same time, bank lending grew from 15 percent to 36 percent of the total.[29] There are a number of reasons that private banks have shouldered greater responsibility over the last decade. On the demand side, first and most obviously, needs have been rapidly growing in Third World countries at precisely the time that official development assistance has diminished. Slow economic growth combined with entitlement pressures in OECD countries have had and can be expected to continue to have a negative impact on foreign assistance. Second, the twin oil shocks created tremendous emergency demand for capital in many less developed countries. On the supply side, private banks were willing and able to respond to these needs because of swollen dollar reserves resulting from a massive flow of petrodollars from the Middle East.[30]

In 1970, private sector lending played a very small role in North–South capital transfers, with bank lending accounting for only three billion dollars. This quadrupled to twelve billion in 1975 and reached thirty-six billion by 1983. The first oil shock created an overwhelming need for dollars to pay inflated oil bills in many of the oil-importing LDCs. At the same time the OPEC banker countries, overwhelmed by a growing flow of dollars, parked many of them in foreign bank accounts, creating both a reservoir of capital that could be lent to LDCs and considerable pressure on the banks to do so. The second oil shock created similar needs and banks were able to respond with $115 billion in loans during the period 1980–83.[31]

The growing repayment problem of the 1970s was transformed into the debt crisis of the 1980s through a series of questionable policy decisions made in the United States.[32] The inflationary tide following the second oil shock drove up energy-related prices in the United States. The Reagan administration and the Federal Reserve, fighting this new-fashioned inflation with old-fashioned remedies, squeezed the domestic money supply, drove up domestic interest rates, and created a deep domestic recession that eventually became global in scope. Unprecedented high real interest rates in the United States both choked off economic growth and elevated the value of the dollar. Other OECD countries followed suit with their own defensive interest rate increases in attempting to slow the flow of capital to the United States, thus spreading the recession around the world.

Large borrowers in the less developed world were thus buffeted in the 1980s by a quadruple shock that drastically changed prospects for capital accumulation. First, terms of trade and trade balances were adversely impacted by the rapidly rising energy prices associated with the second oil shock. Then the Reagan monetary shock drove up interest rates on their rapidly growing privately held debt. For the middle income economies, for example, the average interest rate on outstanding debt jumped from 6.2 percent in 1970 to 10.0 percent in 1984.[33] Furthermore, lenders quickly increased the percentage of loans with floating interest for all less developed countries from 16.2 percent in 1974 to 42.7 percent in 1983, thus tying repayment schedules to financial conditions in the OECD countries.[34] The enhanced value of the dollar, created by the unprecedented high interest rates, further shocked the debtor countries, since about three quarters of the world's long-term debt was denominated in dollars.[35] Finally, after some delay, a fourth shock was felt by most of these countries as the worldwide recession dried up markets for their exported commodities.

Assessing the scope of the resulting full-blown world debt crisis is a complex undertaking. If the gross size of debt is taken as the only indicator of which countries are in trouble, then the United States clearly qualifies as having the most serious problem. The persistent trade deficits of the 1980s quickly transformed the United States into the world's largest net debtor, passing Brazil in 1986. Although concern over the growth of U.S. debt is mounting, the position of the United States in the world economy and the role played by the dollar are factors that make its debt a special case. Table 6-5 lists the other major LDC debtor countries by size of gross foreign debt, public and private, owed to various international creditors. The bulk of private bank lending has gone to what bankers have thought were the most creditworthy countries. Thus, in terms of size of debt, Brazil and Mexico are in the most serious trouble and cause the greatest concern to many American banks exposed to risks of never being repaid.

Other indicators also express the severity of the repayment problem for individual countries. One useful measure of the ability to meet financial commitments is the debt service ratio, the ratio of the principal and interest due annually on loans to the annual revenue received from exports. The debt service ratio can obviously be affected by changes in interest rates as well as by demand for a country's exports, and thus it fluctuates significantly. Conservative World Bank figures indicate that both Brazil and Mexico are in dire straits, given that more than half of export earnings should in theory be devoted to debt repayment.[36] Other data show that Argentina should devote 52 percent of gains from exports to servicing payments, while Chile and Brazil both should pay over 40 percent of similar earnings for debt servicing. Although the low income countries are of less concern to the large private banks, their total exposure being much lower there, many of these impoverished countries

Table 6-5 Major LDC debtors (figures represent billions of dollars).

	1982	1984	1987*	Debt Service Ratio (1985)**
Brazil	85	102	108	35
Mexico	87	95	100	48
Korea	37	44	52	22
Argentina	42	45	50	24 (1983)
Indonesia	25	32	43	25
India	—	31	40	13
Venezuela	36	35	35	15 (1983)
Turkey	22	25	30	32
Philippines	25	27	28	20
Nigeria	14	19	23	32
Chile	18	21	20	44

*Estimates from *The Wall Street Journal* (January 9, 1987) and *The Washington Post* (February 24, 1987).

**Payments of principal and interest as a percentage of exports of goods and services. Although data for them are incomplete, Poland and Iraq rank among the major debtors.

SOURCE: Morgan Guaranty Trust Company, World Financial Markets; World Bank, *World Development Report 1987.*

must also devote more than one quarter of their meager export earnings to making required payments.

The four financial shocks experienced by debtor countries in the 1980s have thus put many of them into an untenable position, akin to national bankruptcy. When financially troubled countries are no longer able to meet repayment obligations, they undergo a process to obtain financial relief called debt rescheduling. Working out the details of a debt rescheduling is a complicated and messy business. Most insolvent countries have a wide variety of obligations to governments, international organizations, and often more than a hundred private banks. Obligations that are primarily of an official nature—money owed to or guaranteed by governments—are negotiated within the framework of the Paris Club. Established in 1956 by a group of creditor countries engaged in bailing out Argentina, the Paris Club is a loose-knit group that meets from time to time to consolidate and reschedule payments on official government-to-government obligations that are beyond the current capabilities of debtor countries. Rescheduling of private sector obligations is handled by an advisory committee called the London Club, which is composed of representatives of creditor banks.

The politics of debt rescheduling have taken intriguing turns over the last decade as the obligations of debtor countries have continued to mount. In the early 1980s, debtor countries unable to meet their public or private commitments came to creditors as supplicants seeking relief. Both creditor clubs normally analyzed the merits of the case, a process that involved scrutinizing the internal as well as external economic situation facing the petitioning country. The Paris Club would establish a framework for handling arrears, and then the debtor country would enter into bilateral agreements with creditors. The London Club operated in a similar manner: an advisory committee, acting in the name of all the private creditors, would outline an acceptable program of relief, a process usually accompanied by an extra interest penalty of about 1½ percent. In both cases the International Monetary Fund (IMF) acted as the financial sheriff, analyzing the economic difficulties of the petitioner countries and prescribing sometimes bitter economic and political remedies as conditions for debt rescheduling.[37]

The IMF is an arm of the World Bank that has historically provided short-term funding for countries with adjustment difficulties. It has traditionally been dominated by orthodox economic thinking anchored in free-market philosophies. The conditions prescribed for debt rescheduling have usually reflected this perspective. IMF conditionality normally required LDC debtors to undertake austerity programs, which would include major cutbacks in government spending and subsidies as well as currency revaluations designed to make exports more attractive.[38] Often these IMF programs have been economically sound but politically naive. Introduction of IMF austerity programs has often been accompanied by political unrest, protest, and even rioting in the debtor countries. More recently, the IMF has become somewhat more politically sensitive in setting the conditions by which it, and indirectly the Paris and London clubs, will reschedule debt.

As obligations continued to grow in the face of world economic recession in the 1980s, debtor countries became more critical of the IMF and the approach taken by commercial banks. Small countries like Peru suggested forming a debtors' cartel, an organization of countries that would jointly refuse to meet repayment obligations. The large debtors such as Brazil and Mexico, however, being dependent on the future goodwill of the banking community, refused to go along with the initiative. In a sense, being a small debtor gives an individual country very little economic and political clout. But being a large debtor can change the situation considerably, since banks are likely to be more accommodating when much more is at stake.

An international game of financial chicken played between banks and debtors culminated in July of 1985 when newly elected president Alan Garcia of Peru declared that, because of pressing internal development needs, only 10 percent of Peru's export revenue would henceforth be available for debt servicing. Astonished bankers could only respond with refusals to make further credit available, since they had no troups to send to seize Peruvian assets. Emboldened by Garcia's actions and burdened by their own severe debt situations, both Mexico and Brazil subsequently declared their own independence from the IMF and private banks. In February 1986, Mexican president Miguel de la Madrid Hurtado declared that Mexico would limit its debt repayments according to its capacity to repay, a politely veiled way of calling for further negotiations on the definition of Mexico's ability to pay. The U.S. government reportedly responded with maneuvers behind the scenes, came up with a generous rescheduling agreement at least partially pegged to Mexican growth rates, and presented it to creditors as a fait accompli.[39] Then in early 1987, culminating more than two years of fencing with the IMF, Brazil flexed its muscles by demanding the same kind of treatment accorded Mexico. In this case, the private banks, which were collectively owed more than sixty-seven billion dollars, were less willing to accede to Brazil's demands.

The politics of the debt trap will undoubtedly become more complicated over the next decade. The banks are far from united in their approach to debt rescheduling, particularly in the case of Brazil. The large number of banks involved have different interests in proposed settlements based on the amount of exposure. Small regional banks, for example, may have only modest exposure and seek to remove themselves from an agreement entirely by having a bigger bank take over their loans. Such loan transfers free the small banks of the danger of loss and eliminate possible vetoes from rescheduling proceedings. Large banks with significant exposure, however, have a different set of interests. Citicorp loans to Brazil amounted to more than $5 billion in 1987, and the bank stood to lose $340 million in yearly income in a default. Other banks with somewhat smaller exposure stood to lose as much as 32 percent of annual profits and would be forced to write down assets considerably because of the nonperforming loans.[40]

The problems are complex and the stakes are high for all participants in the international debt skirmishing. Lender countries are concerned that large-scale LDC defaults could severely damage or destroy many large banks, leading to a domino effect in bank failures that could only be stemmed by heavy government involvement. The major debtors are concerned because their fluctuating export earnings often fail to produce enough foreign exchange to meet financial obligations. Thus, Mexico was able to negotiate assistance from the IMF based on the price of a barrel of oil, and Argentina has attempted to get automatic lending increases tied to the price of a bushel of grain.[41] In addition, since most new loans are at floating interest rates tied to either the LIBOR (London Inter-Bank Offered Rate) or the U.S. prime rate, monetary policy in the industrial countries is now constrained by the need to keep interest rates low enough to avoid a major debt repayment catastrophe. Thus, it would now be impossible to squeeze the U.S. money supply and drive up interest rates to 1981 levels without bringing on dramatic and large-scale debtor country defaults.

Politics in the debtor countries also constrain future possibilities. International borrowing has been an essential source of capital to maintain the stability of many LDCs with rapidly growing populations and related unemployment. Most of these major debtor countries have been living well beyond their means, but belt tightening is a politically difficult process that threatens both political stability and the fragile democracies that have evolved in many of these countries. The problem is compounded by capital flight. Lacking confidence in the future, many of the politically powerful and economically affluent in these debtor countries have squirreled away large chunks of capital in foreign bank accounts. It is estimated that between 1976 and 1985 about $123 billion fled from ten key Latin American debtor countries and $75 billion fled from eight key debtors in other parts of the world. In the period 1979–82, $26.5 billion fled from Mexico, $22 billion from Venezuela, and $19.2 billion from Argentina. In each case, the amount of capital flight represented much more than 50 percent of new capital inflows.[42] U.S. banks, which together hold approximately $83 billion in Latin American loans, are now thought to hold a similar amount of capital in private bank accounts on behalf of Latin American residents living in these debtor countries.[43]

▼ Toward Sustainable Modernization

The many factors limiting prospects for countries on the other side of the development gap add up to a complex and somewhat pessimistic picture. The colonial legacy in most of the low income countries is still to be overcome. Ecological and technological factors have combined to lessen opportunities for newcomers in the global economy. Incomplete modernization in the low income countries has led to a population explosion that makes per capita economic growth difficult. The pace of technological innovation in the already developed countries helps perpetuate the gap between rich and poor countries. There are few opportunities for low income economies to develop competitiveness in anything but raw materials, and it is difficult for LDCs to compete with the technological giants that now dominate world markets. Official development assistance and private capital flows to most of these countries are not increasing in real terms because of budget crises in donor countries and the inability of potential recipients to service existing debt. Even the values governing industrialization processes have changed dramatically, placing greater obstacles in the way of contemporary modernization efforts.

The apparent failure of many less developed countries to make substantial progress in closing the gap represents a significant anomaly in the liberal modernization model. Open markets and free trade haven't solved development dilemmas, and the various prescriptions for growth offered by development experts have had only mixed successes at best.[44] A few countries, particularly the Asian NICs, have made progress, but largely because of extensive aid and good timing for their export-oriented growth strategies.[45] And the slow growth of world markets over the last decade has made it impossible for second-tier NICs to emulate these successes.

There are no simple prescriptions for rapidly transforming the low income countries. Learned commissions have met, deliberated, and suggested a number of steps that could be taken to close the income gap.[46] But idealistic prescriptions are rarely implemented. Part of the solution must involve more careful and honest analysis of the dilemmas of the least developed countries. Benign neglect doesn't serve the interests of countries on either side of the gap. In an ecologically interdependent world, the squalor and misery endemic to the least developed countries translates into potential worldwide plagues, such as AIDS, that jump quickly from one country to another. Terrorism and related political violence also breed in countries where there is little hope that playing by conventional rules will lead to better living standards. And from a liberal value perspective, it is increasingly difficult to live in a global village badly divided between the wealthy and those trapped in poverty.

A new vision, or paradigm, for dealing with the problems of development is very much needed. Such a vision must take account of the diversity of development problems in the Third World and suggest sets of tasks for countries on both sides of the gap. For most of the least developed countries, controlling burgeoning populations is clearly a primary concern. But until Third World politicians are willing to confront traditional elements that favor large families, very little of value can be accomplished. Internal feuding and conflict in LDCs must also be brought to a halt. Most of the poorest economic performers are countries that have been beset by various forms of civil strife, and little can be done to enhance the quality of life until these conflicts are eliminated.

On the other side of the gap, the technologically developed world must take responsibility for maintaining an adequate flow of technology and capital to the less developed world. Aid should be given to countries that can make best use of it and not only to strategic allies. The heavy stress on military aid must be eliminated if any significant change in economic growth prospects is to take place. Furthermore, political leaders in the countries that industrialized early must accept the fact that successful development of the Third World can only take place at some expense to their own interests. There is little possibility that the liberal promise of a more equitable distribution of global growth will be fulfilled without significant adjustment problems arising in domestic markets.[47] Finally, research and development that concentrates on problems of less developed countries should be accelerated so that technology begins to reduce, rather than to accentuate, problems of global inequality.

Even if major efforts are made by all involved countries to close the development gap, it will never be possible for the greatest share of humanity to live at the levels of material consumption reached by the early industrializers.[48] There simply isn't enough potential resource capacity to support material-intensive industrialization on a planetary scale. This shouldn't be taken to mean that the situation is hopeless. On the one hand, third-wave post-industrial growth is stabilizing or reducing raw material consumption in the industrialized world.[49] And on the other hand, countries like China are studying possibilities for modernization without industrialization, possibly skipping the material-intensive second-wave growth and exploiting third-wave technologies. Modernization of education, mass communications, basic medical care, and agriculture may well lead to a better quality of life in many of the least developed countries without exposing citizens to the pollution, dislocation, and turmoil associated with heavy industrialization and urbanization. Such a new philosophy of development suggests that certain countries may skip the agonies of the industrial revolution and move beyond the centralization and materialism inherent in the contemporary industrial paradigm.

A new development paradigm cannot be constructed overnight, but continued exposure of anomalies in the old one argues for more realistic assessments of development prospects. Post-industrial development should be built around people and their needs rather than mobilizing them in support of shopworn industrialization strategies. A new ethics of development must be created to replace the rampant materialism that is now sold to the less developed world under the guise of progress.[50] This process of designing a new development model could well result in a more sustainable form of modernization, characterized by higher levels of human satisfaction while avoiding continuation of an all-out, destructive attack on the global ecosystem.

▼ Notes

1. See, for example, Samir Amin, *Unequal Development: An Essay on the Social Formation of Peripheral Capitalism* (New York: Monthly Review Press, 1976); Arghiri Emmanuel, *Unequal Exchange: A Study of the Imperialism of Trade* (New York: Monthly Review Press, 1972); Fernando Cardosa and Enzo Faletto, *Dependency and Development in Latin America* (Berkeley: University of California Press, 1979).
2. See John Lewis and Valeriana Kallab, ed., *Development Strategies Reconsidered* (New Brunswick, N.J.: Transaction Books, 1986).

3. Louis Emmerij, "The Distribution of Income Among Nations," *The OECD Observer* (November 1986).
4. Eric Larson et al., "Beyond the Era of Materials," *Scientific American* (June 1986).
5. World Bank, *World Development Report* 1986 (New York: Oxford University Press, 1986), table 28.
6. For an overview of these problems see Jacques Loup, *Can the Third World Survive?* (Baltimore: Johns Hopkins University Press, 1983).
7. See Marian Radetzki, "Where Should Developing Countries' Minerals be Processed? The Country View Versus the Multinational Company View," *World Development* (April 1977); Marian Radetzki, "Has Political Risk Scared Mineral Investments away from the Deposits in Developing Countries?" *World Development* (July 1982).
8. See Mark Zacher and Jock Finlayson, *Developing Countries and the Commodity Trading Regime* (New York: Columbia University Press, 1988); Mark Zacher, "Trade Gaps, Analytical Gaps: Regime Analysis and International Commodity Trade Regulation," *International Organization* (Spring 1987).
9. For an analysis of problems associated with the Lomé Convention, see Adrian Hewitt, "Stabex: An Evaluation of the Economic Impact over the First Five Years," *World Development* (December 1983).
10. See Robert Clark, Jr., *Development and Instability* (New York: Dryden Press, 1974), chap. 8.
11. See Erik Eckholm, *The Other Energy Crisis: Firewood* (Washington, D.C.: Worldwatch Institute, 1975); Sandra Postel and Lori Heise, "Reforesting the Earth," in Lester Brown et al., *State of the World 1988* (Washington, D.C.: Worldwatch Institute, 1988).
12. See Eduardo Lachica, "U.S. Asks World Bank to Make Safeguarding Environment a Priority," *The Wall Street Journal* (July 3, 1987); Eduardo Lachica, "Fear of Environmental Damage Delays Development Projects in Third World," *The Wall Street Journal* (March 16, 1987).
13. Mahbub Ul Haq, *The Poverty Curtain* (New York: Columbia University Press, 1976), p. 20.
14. William Kline, "Can the East Asian Model of Development be Generalized?" *World Development* (February 1982).
15. See Stephan Haggard and Chung-In Moon, "The South Korean State in the International Economy: Liberal, Dependent or Mercantile?" in John Ruggie, ed., *The Antinomies of Interdependence* (New York: Columbia University Press, 1983).
16. See Susan Strange and Roger Tooze, eds., *The International Politics of Surplus Capacity* (London: Butterworth, 1980); "Glutted Markets: A Global Overcapacity Hurts Many Industries," *The Wall Street Journal* (March 9, 1987).
17. Data from *Handbook of International Trade and Development Statistics 1986 Supplement* (New York: United Nations, 1986), table 4.11.
18. Ibid., table 2.7a.
19. See "Want Some O.J.? It's Fresh from the Test Tube," *Business Week* (November 17, 1986).
20. Stuart Auerbach and Ward Sinclair, "U.S. Cuts Quotas for Sugar," *The Washington Post* (December 16, 1986).
21. See LaMond Tullis, "Cocaine and Food," paper presented to the Annual Convention of the International Studies Association (March 1986); "Can South America's Addict Economies Ever Break Free?" *Business Week* (September 22, 1986).
22. World Bank, *World Development Report 1985* (New York: Oxford University Press, 1985), p. 101.
23. An excellent analysis of humanitarian motives in foreign aid is found in Roger Riddell, *Foreign Aid Reconsidered* (Baltimore: Johns Hopkins University Press, 1987), Part I.
24. Data are from World Bank, *World Development Report 1986*, tables 20, 21.

25. U.S. Department of State, *Foreign Assistance Program: FY 1986 Budget and 1985 Supplemental Request* (Washington, D.C.: May 1985), p. 10.
26. Mark Valentine, "Foreign Assistance Working Group: FAWG Recommends to Fight Aid Cuts," *Interaction* (August 1986).
27. See, for example, Richard Barnet and Ronald Muller, *Global Reach: The Power of the Multinational Corporations* (New York: Simon & Schuster, 1974).
28. World Bank, *World Development Report 1985*, p. 4.
29. Ibid., pp. 19–21.
30. Alfred Watkins, *Til Debt Do Us Part* (Lanham, Md.: University Press of America, 1986), chap. 2.
31. World Bank, *World Development Report 1985*, p. 21.
32. See John Makin, *The Global Debt Crisis* (New York: Basic Books, 1984), Part III; Miles Kahler, "Politics and International Debt: Explaining the Crisis," *International Organization* (Summer 1985).
33. World Bank, *World Development Report 1986*, table 19.
34. World Bank, *World Development Report 1985*, p. 21.
35. Ibid., p. 22.
36. World Bank, *World Development Report 1986*, table 17.
37. See Anthony Sampson, *The Money Lenders* (New York: Penguin, 1981), chap. 21.
38. See Stephan Haggard, "The Politics of Adjustment: Lessons from the IMF's Extended Fund Facility," *International Organization* (Summer 1985).
39. Peter Truell, "Brazil Will Press for Concessions Won by Mexico," *The Wall Street Journal* (January 9, 1987); see also "The Debtor's Revolt is Spreading in Latin America," *Business Week* (December 28, 1987).
40. Peter Truell, "Citicorp Signals Harder Stance on Brazil Debt," *The Wall Street Journal* (March 16, 1987); "Brazil and Its Creditors: Who Has More to Lose?" *Business Week* (March 9, 1987).
41. James Rowe, Jr., "Argentina Seeks New Lending Plan," *The Washington Post* (September 5, 1986); see also "LDC Debt: Debt Relief or Market Solutions?" *World Financial Markets* (September 1986).
42. World Bank, *World Development Report 1985*, p. 64.
43. See James Henry, "Where the Money Went," *The New Republic* (April 14, 1986); "Has Capital Flight Made the U.S. a Debtor of Latin America?" *Business Week* (April 21, 1986).
44. See John Lewis, "Development Promotion: A Time for Regrouping," in John Lewis and Valeriana Kallab, eds., *Development Strategies Reconsidered.*
45. See Colin Bradford, Jr., "East Asian 'Models': Myths and Lessons," in John Lewis and Valeriana Kallab, eds., op. cit.
46. See, for example, Commission on International Development Issues, *North–South: A Program for Survival* (Cambridge, Mass.: MIT Press, 1980); World Commission on Environment and Development, *Our Common Future* (New York: Oxford University Press, 1987).
47. See John Culbertson, "Destructive Foreign Trade: Sowing the Seeds of Our Own Downfall," *The Futurist* (November–December 1986).
48. For example, should China's standard of living be elevated to the U.S. level, the result would be ecological disaster. These calculations were carried out in Dennis Pirages and Paul Ehrlich, "If All Chinese Had Wheels," *The New York Times* (March 16, 1972).
49. See Alvin Toffler, *The Third Wave* (New York: Morrow, 1980), chap. 23; Eric Larson et al., op. cit.
50. See Dennis Goulet, *The Uncertain Promise: Value Conflicts in Technology Transfer* (New York: IDOC, 1976).

▼ Suggested Reading

ROBERT AYRES, *Banking on the Poor* (Cambridge, Mass.: MIT Press, 1983).

CHRISTOPHER BROWN, *The Political and Social Economy of Commodity Control* (New York: Praeger, 1980).

CHRIS CARVOUNIS, *The Foreign Debt National Development Conflict* (Westport, Conn.: Quorum Books, 1986).

FERNANDO CARDOSA AND ENZO FALETTO, *Dependency and Development in Latin America* (Berkeley: University of California Press, 1979).

ROBERT CASSEN ET AL., eds., *Rich Country Interests and Third World Development* (London: Croom Helm, 1982).

BENJAMIN COHEN, *In Whose Interest? International Banking and American Foreign Policy* (New Haven, Conn.: Yale University Press, 1986).

PETER EVANS, *Dependent Development* (Princeton, N.J.: Princeton University Press, 1979).

RICHARD FEINBERG AND VALERIANA KALLAB, eds., *Adjustment Crisis in the Third World* (New Brunswick, N.J.: Transaction Books, 1984).

RICHARD FEINBERG AND VALERIANA KALLAB, eds., *Uncertain Future: Commercial Banks and the Third World* (New Brunswick, N.J.: Transaction Books, 1984).

PRADIP GHOSH, ed., *Technology Policy and Development: A Third World Perspective* (Westport, Conn.: Greenwood Press, 1984).

DENNIS GOULET, *The Uncertain Promise: Value Conflicts in Technology Transfer* (New York: IDOC, 1976).

MILES KAHLER, ed., *The Politics of International Debt* (Ithaca, N.Y.: Cornell University Press, 1986).

DONALD LESSARD AND JOHN WILLIAMSON, eds., *Capital Flight and Third World Debt* (Washington: Institute for International Economics, 1987).

JOHN P. LEWIS AND VALERIANA KALLAB, eds., *Development Strategies Reconsidered* (New Brunswick, N.J.: Transaction Books, 1986).

JACQUES LOUP, *Can the Third World Survive?* (Baltimore, Md.: Johns Hopkins University Press, 1983).

JOHN MAKIN, *The Global Debt Crisis* (New York: Basic Books, 1984).

MORRIS MILLER, *Coping is Not Enough* (Homewood, Ill.: Dow Jones–Irwin, 1986).

North-South: A Program for Survival (Cambridge, Mass: MIT Press, 1980).

ERNEST PREEG, ed., *Hard Bargaining Ahead: U.S. Trade Policy and Developing Countries* (New Brunswick, N.J.: Transaction Books, 1985).

JOHN RAVENHILL, *Collective Clientelism: The Lomé Convention and North-South Relations* (New York; Columbia University Press, 1985).

MICHAEL REDCLIFT, *Sustainable Development: Exploring the Contradiction* (New York: Methuen, 1987).

PAUL REYNOLDS, *International Commodity Agreements and the Common Fund* (New York: Praeger, 1978).

ROGER RIDDELL, *Foreign Aid Reconsidered* (Baltimore, Md.: Johns Hopkins University Press, 1987).

ANTHONY SAMPSON, *The Moneylenders* (New York: Penguin, 1981).

SCOTT SIDELL, *The IMF and Third-World Instability* (New York: St. Martin's Press, 1988).

PAUL STEIDLMEIER, *The Paradox of Poverty: A Reappraisal of Economic Development Policy* (Cambridge, Mass.: Ballinger, 1987).

FRANCES STEWART, *Technology and Underdevelopment* (Boulder, Colo.: Westview Press, 1977).

PAUL STREETEN, ed., *First Things First: Meeting Basic Human Needs in Developing Countries* (New York: Oxford University Press, 1981).

ALFRED WATKINS, *Til Debt Do Us Part* (Lanham, Md.: University Press of America, 1986).

MARK ZACHER AND JOCK FINLAYSON, *Developing Countries and the Commodity Trading Regime* (New York: Columbia University Press, 1988).

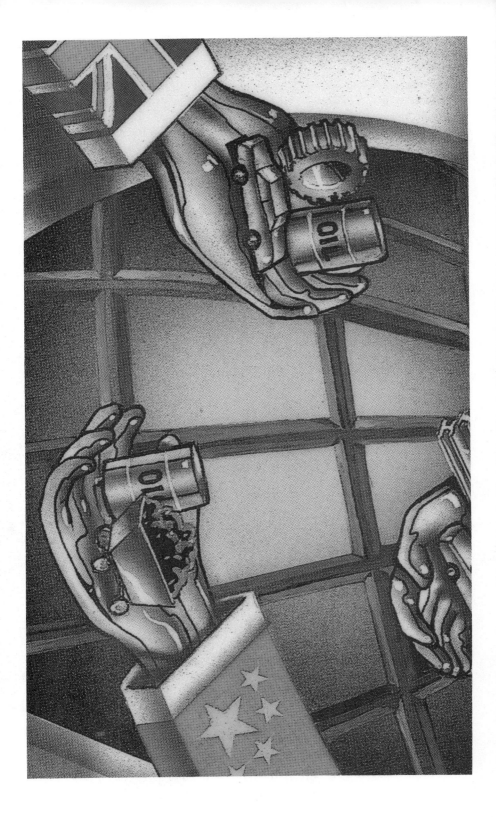

7

▼

The
Politics of
Technology
Diffusion

Changing patterns of technological innovation and related differential growth rates and trade balances are now a major source of friction in relations among the industrialized market economies. Following World War II, the United States emerged as the dominant authority, or hegemon, in international politics and economics. But successful postwar U.S. international economic policy resulted in a resurgent Western Europe and Japan, both of which have emerged as major challengers to postwar U.S. domination. Intensified high-technology trade interaction has moved from competition to potential conflict as Japan's economic miracle has created massive trade deficits for the United States and led to U.S. demands for new economic rules to create a "level playing field" for all countries.

As the hegemony of the United States has faded and relations among the industrial powers have become more consensual, the understood rules of good conduct in economic matters have been called into question. In such situations where there is no clear authority, individuals and nations are more free to pursue individual interests, often at the expense of the collective. This type of behavior need not be motivated by evil intent but can result from an incomplete assessment of the impact particular actions will have. Situations in which pursuit of self-interest is immediately rewarded but leads to long-term individual or collective harm are called *social traps*.[1] There are many varieties of these traps. Time delay traps, for example, occur when behavior is rewarding in the short term but leads to undesired consequences in the long term. Cigarette smoking is a case in point. Collective traps occur when individual actions are not harmful in isolation but when combined lead to collective trouble. An example would be burning fireplace logs in a densely populated ski resort leading to foul air in a once pristine environment. In other contexts, these social traps are referred to as commons tragedies or problems of coordination.[2]

Leaders of nations, like other individuals, often engage in realpolitik behavior that yields immediate and obvious political rewards but may create long-term domestic and global problems. An example of such a domestic situation is the massive U.S. deficit spending of the 1980s, whereby politicians bought short-term prosperity by running up a massive national debt that greatly restricts options for future generations. An example of the global situation is the inexorable buildup of carbon dioxide in the atmosphere as a result of the expansion of polluting industrial activity in many nations around the world. In the short run energy-intensive industrialization keeps the world's economies humming, but in the long run the carbon dioxide buildup is altering world climate.

International social traps can assume many different shapes. In the ecological domain they tend to be collective traps, or commons tragedies. Ecologist Garrett Hardin has popularized the commons tragedy as a metaphor that applies to contemporary global ecological problems.[3] Examples of social traps, such as the global carbon dioxide buildup, are easy to spot in the ecological realm; but there are other types of less obvious international social traps now being baited in the structural realm. These are often time delay traps and coordination problems. They are set when the activities of many individual countries may create an eventual problem for the international economy, but the short-term rewards are so obvious and compelling as to override any long-term thinking. The current wasteful duplication of production facilities in many industrial and industrializing countries, which is creating eventual problems of surplus capacity, is an example of a time trap and coordination problem. Individual countries pursuing their own national interests build redundant production capacity in many industries, thus wasting scarce capital on what shortly become idle production facilities in overbuilt industries.

Three sets of dilemmas or paradoxes are associated with these various social traps; they result from the ongoing transformation of the world into a global village in which there is no superordinate authority to solve novel and complex problems of coordination. Foremost among these dilemmas is increasing system interdependence in the face of domestic demands for greater autonomy. In the years following World War II, large oceans made good neighbors of the industrial countries. But changes in transportation and communications have served to shrink the physical and psychological distances among countries. The industrialized OECD countries composing the Western Alliance have had their share of political and economic misunderstandings over the last few decades, but their relative autonomy, widespread economic growth, and perceived common enemy (the Soviet Union) served to keep such friction from degenerating into conflict. Now, continuous advances in science and technology are eroding national borders, and independent nations in the industrial section of the emerging global village find their freedom increasingly constrained by new types of ecological, structural, policy, and value interdependence.

A second technology-driven set of paradoxes that bedevils OECD countries is an emerging equality among the major powers in the face of a greater need for a hegemonic power. In the aftermath of World War II, the United States emerged as the unquestioned scientific, technological, economic, and political power in the international system. The rules of the postwar political and economic order were largely determined by the United States in the absence of a challenge from any other country. The economies of the wartime enemies, Japan and Germany, were destroyed by the hostilities, and leaders in these countries had very little voice in structuring the postwar international system. The French, British, and Russians were also in dire economic straits and raised few objections to the postwar order crafted and dominated by the United States.

The United States used its technological, military, and economic edge to create a postwar order that brought increasing prosperity to those willing to play by its rules. But ironically, U.S. policies during this period were gradually sowing the seeds that would undermine the postwar industrial order the United States had set up.[4] The Marshall Plan and related aid programs were very successful in meeting immediate goals of rebuilding the political and economic systems of former enemy and ally alike. But the resulting system, a product of the successes of U.S. postwar policy, is currently in flux as U.S. leadership is no longer uncritically accepted by the other major powers, which are seeking to carve out a more influential role for themselves.

Both increased interdependence and the loss of U.S. hegemony have been instrumental in spawning significant trade friction between the United States and its industrial partners. The rules of international commerce, largely shaped by free-market doctrines, have been determined by the United States. By default, the dollar has been a sort of universal currency to which others have been pegged. Under these circumstances, leaders of U.S. industry have worried little about foreign competition and even less about penetrating foreign markets. Indeed, the critical literature of the 1960s and 1970s complained about the global reach of U.S.-based multinational corporations and U.S. dominance in global economic affairs.[5] But the Japanese, Germans, and others have honed their competitive skills and developed export strategies to capture larger market shares. Over the last decade, the Japanese in particular have been highly successful in flexing their economic muscles to the detriment of the United States. Facing recent annual trade deficits in excess of $160 billion, U.S. spokespersons have increasingly decried what they perceive to be unfair trade tactics used by others, and peaceful trade competition threatens to degenerate into trade warfare. In the 1990s the shoe may be on the other foot as the global reach of foreign corporations expands and investors snap up larger portfolios of U.S. assets.

While West–West relations have become contentious because of these developments, East–West relations are also being reshaped by techno-ecological factors. The Soviet Union is emerging as a full-fledged member of the industrial fraternity. While the USSR does not yet rival the United States or Japan in technological prowess across the board, in certain areas such as fusion research and space exploration Soviet achievements are certainly as impressive as those of other industrial countries. Soviet leaders see better relations with the West as a way of obtaining technology and capital in order to accelerate Soviet economic growth. In return, the Soviets hope to trade their vast quantities of natural resources, including abundant supplies of petroleum and natural gas, with Western European resource-deficient countries. There is thus a potential natural partnership between the Soviet Union, in need of capital and technology, and Japan and Western Europe, which could make good use of Soviet natural resources.

Hard-liners in the United States have been strongly opposed to any significant economic dealings with or transfer of technology to the Soviet Union. U.S. leaders have disagreed with European allies over the desirability of commerce with the Russians. Recent internal political changes in the Soviet Union have intensified arguments within the United States and the Western Alliance over the question of doing business with the Soviets. A paradox here is that those countries willing to cut a deal with the Soviet Union will certainly profit in the short run, though they could be transferring technologies that might sow the seeds of destruction for the Western Alliance in the long term.

A third set of technology-related dilemmas revolves around more intense competition among the OECD countries for LDC export markets in the face of rapid development of fundamentally transforming and potentially dangerous cutting-edge technologies. This combination is creating unprecedented North–South technology transfer issues. In particular, new innovations in nuclear power and in biotechnology raise significant questions about the appropriateness of diffusion of such know-how to politically unstable countries. The transfer of fuel enrichment and reprocessing technologies for nuclear power paves the way for recipient countries to fabricate

nuclear weapons should they desire to do so. Biotechnology raises a real possibility for the creation and release of harmful new organisms, intentionally or unintentionally, which could devastate existing ecosystems.

▼ National Autonomy and Global Interdependence

The acceleration of scientific discovery and technological innovation has been accompanied by the growth of various kinds of interdependence among the industrial market economies that diminish possibilities for maintaining traditional types of national autonomy. Foremost among these new forms of interdependence is greater ecological interdependence brought on by the considerable raw material requirements of the ongoing industrial revolution. As the industrial revolution has moved forward, the industrial countries, with the notable exception of the Soviet Union, have become increasingly dependent on others to supply the fuels and minerals required to sustain contemporary levels of living. Not only has this made them vulnerable to possible machinations of resource suppliers, but it has also intensified resource competition among them. For example, U.S. oil companies began importing significant amounts of petroleum from the Middle East in the late 1960s and early 1970s, a move that was frowned on by other OECD oil importers and that was at least partly responsible for creating the tight petroleum market that preceded the first oil crisis of 1973–74.

Table 7-1 indicates the depth of OECD country dependence on foreign suppliers of fuels, minerals, and other raw materials. Japan, with only about 3 percent of the world's population, is clearly in the most precarious situation with nearly one hundred billion dollars of primary product imports per year. Japanese financial burdens due to imported raw materials peaked in 1980 along with the prices of imported fuels and then diminished somewhat as commodity prices dropped as a result of the global economic recession. But Japanese leaders are well aware of the potential economic impact of another energy crisis, and they shape their foreign and domestic policies accordingly. The members of the European Economic Community (EEC) have experienced the growth of similar ecological pressures, and their combined primary

Table 7-1 Trade balance for basic commodities, 1986 (figures represent billions of dollars).

	Food	Raw materials	Fuels
Japan	−19.3	−17.7	−35.3
United States	−.9	−3.7	−31.6
West Germany	−10.1	−8.0	−18.0
France	4.4	−2.8	−13.0
Italy	−8.1	−6.9	−14.6
United Kingdom	−7.3	−3.9	3.5
Canada	2.9	11.0	4.4

SOURCE: Central Intelligence Agency, *Handbook of Economic Statistics 1987.*

product imports are now larger than those of Japan. The major exception to the pattern is in agriculture, where generous domestic subsidies in EEC countries have reduced food imports. The United States and Canada are generally more self-sufficient than the other industrial market economies. Canada remains an exporter of primary products, while the United States is mainly deficient in fuels.

This growing ecological web creates several kinds of issues in relations among the industrial countries. First, these big trade deficits in primary commodities must be offset by exports of manufactured goods. Thus, Japan has flooded world markets with consumer and high-technology products, often at the expense of good relations with other industrial countries. Second, the quest for natural resources leads to potential conflicts among countries seeking secure access to needed raw materials. When the United States began importing substantial quantities of petroleum from the Middle East in the 1970s, it caused friction with countries that already had resource interests there. The Japanese, having built a natural-resource sphere of influence in the Far East, have now begun moving into Latin America with a plan to gain influence there while helping to reduce the region's debt burden.[6] Third, attempts have been made to reduce primary product dependence through generous domestic subsidies to agricultural interests—an economic policy that not only denies potential markets to exporters of agricultural commodities, but also creates added conflict when surpluses thus generated are dumped on world markets at subsidized prices.[7]

The industrial countries are also becoming much more structurally interdependent, their economic and political fortunes rising and falling together. Enhanced transportation and communications capabilities have created integrated world markets and a related, more complex, worldwide division of labor. Multinational corporations now move facilities around the world to take advantage of the best production opportunities. Even Japanese multinationals have built factories in the United States, not only to penetrate the U.S. market, but also in some cases to export products back to Japan. And American multinationals located in Taiwan significantly distort bilateral trade figures by exporting large quantities of products back to the United States.[8] Investment capital now flows across national borders in search of the highest safe rate of return. But as trade barriers have fallen and as fixed exchange rates have given way to market forces, the values of key currencies have rapidly fluctuated, often creating major adjustment problems for the affected countries.

One important result of this growing structural interdependence is that the economic, political, and even social barriers among industrial countries are rapidly falling; people, money, goods, and services are moving across borders in large quantities. This in turn means that mutual concern among the OECD countries about political, economic, and social conditions in other countries has grown. The 1987 stock market collapse and related insecurity in the United States, for example, was quickly transmitted to markets in other industrial countries. Japan's internal industrial policies and related exceptionally large trade balance have been of deep concern in the United States, where growing international indebtedness has become a major argument for protectionist politics. On the other hand, a significant portion of the growing U.S. public debt has been financed by Japanese investors, and even the U.S. stock

market is significantly influenced by Japanese foreign investment decisions. Thus, in the emerging world of the twenty-first century structural interdependence will make it extremely difficult for any market economy to insulate itself from the activities of others, and it will be imperative for leaders to view policy options in global political-economic terms.

Growing structural interdependence among the industrialized countries is mirrored in the development of a similar policy interdependence. Prior to the mid 1970s, political processes in industrial countries took place in relative isolation. U.S. farm policy, for example, was forged by the interplay of domestic interests, with little attention paid to agriculture in other parts of the world. Domestic problems could be corrected by domestic solutions, and policy makers didn't have to worry about perturbations from other countries. Now, increased policy interdependence introduces new complexity into what once was the domain of solely domestic decision making.

Policy interdependence can intrude into domestic affairs in three ways. First, changes taking place in the larger international system can narrow domestic policy alternatives. The two global energy crises, for example, forced oil-importing industrial countries to adopt energy conservation policies. Second, there is now a much tighter link between foreign and domestic policies. The decision of the Carter administration to embargo grain shipments to the Soviet Union in reprisal for the invasion of Afghanistan had such negative repercussions for U.S. farmers that even hard-line Ronald Reagan was forced to rethink the policy. Finally and most important, the domestic policies of one industrial country increasingly have unforeseen impacts on other countries, and they must be carefully analyzed to avoid unintended consequences. Monetary and fiscal policies of the major powers, for example, must be closely coordinated in order to have the desired domestic impacts. It does little good for the Federal Reserve in the United States to fight inflation by tightening the domestic money supply if higher interest rates lead to a big increase in the flow of capital from other countries to the United States. This type of complexity is illustrated by the distortions in international trade that took place in the mid 1980s, when tight monetary policies in the United States caused the value of the dollar to soar. In order to dampen inflation and attract capital necessary to finance a rapidly growing national debt, real interest rates were kept high in the United States, thereby enhancing the value of the dollar but worsening U.S. trade competitiveness. Subsequent lower interest rates in the United States and the declining value of the dollar led to a reverse flow of capital, making domestic debt management more difficult.

The international repercussions of domestic policies are highlighted by arguments about a "level playing field" in international trade. Although quotas, tariffs, and related cross-border restraints have been recognized as impediments to free trade for a long time, more intense trade competition in the 1980s has focused attention on other ways domestic policies are used to manipulate imports and exports. Macroeconomic policies, including tax and interest rates, special tax incentives, and so forth, can be very important in gaining an edge in trade competition. Similarly, various kinds of subsidies given to certain industries and agriculture in response to domestic political pressures can strongly enhance the competitive position of the recipients. And even measures ostensibly taken to protect the public, such as pesticide standards for imported produce or specification of performance standards, can be used to exclude foreign products from domestic markets.[9]

Finally, there are signs that value interdependence—a tendency toward universally accepted sets of standards of right conduct or basic human rights—is developing among the industrial countries. A nascent consensus on such standards is evident in documents like the Universal Declaration of Human Rights, in the Helsinki accords, and in related efforts to define this growing agreement. And although some of the human rights pressure on the Soviet Union in the 1980s could be ascribed to U.S. "bear baiting," the community of industrialized nations obviously feels that a more interdependent world requires a consensus on values and behavior deemed acceptable by the global community.

▼ Increased Competition for Limited Markets

Another source of tension among the industrialized countries and between them and the newly industrializing nations is related to the rapid diffusion of science and technology and the related growth of excess manufacturing capacity. This wasteful situation, too much growth in capacity and too few markets, results from a mercantilist urge to increase national exports at the expense of competitors. A healthy trade balance is considered an important aspect of national security in many countries; and in the less developed countries, export expansion is vital if hard-currency debts are to be serviced. But export growth that may seem optimal to individual countries in the short run can create time delay traps and coordination problems in the long term.

At the end of World War II, the United States assumed unquestioned world leadership and laid out a postwar trade system based on free-market ideals. Still smarting from the global economic depression and the following wartime destruction, liberal economists in the United States prescribed a free-trade system as optimal for spurring international economic growth and thus providing for worldwide economic security. They assumed, perhaps in error, that protectionist policies were responsible for the Great Depression and the war that followed it.[10] According to their view, the less government intervention in restricting the flow of trade, the better off all nations would be. Free trade, to their way of thinking, allowed for an optimal allocation of world resources in the production of goods and services and thus served both global and national interests—at least the national interests of the economically strong nations. Protectionism, they argued, leads to inefficient production and eventually economic losses for all involved. It is by no means coincidental that the free-trade philosophy espoused by U.S. policy makers was ideally suited to maintaining the dominance of U.S. industries in international markets.

This free-trade dogma guided U.S. international economic policy over four decades and only recently has been questioned seriously in the face of strengthened competition from other countries. As long as the United States remained the world's largest creditor nation and the global reach of U.S. corporations continued to expand, liberal trade orthodoxy predominated. But the disappointing trade figures of the late 1980s and the descent of the United States to the position of biggest international debtor have sharpened the argument between free traders and protectionists. Now protectionist schemes once advanced mainly by critics of U.S. policies living in less developed countries are touted as alternatives in the United States.

Approaching the turn of the century, the United States finds itself in a new and uncomfortable position as one competitor among many for economic, political, and military preeminence. In some respects the relative decline of the United States was an inevitable result of what has been referred to as the "hegemon's dilemma," the adoption of economic policies that may be required to maintain the international system but that eventually undermine the dominant position of the hegemonic power.[11] In order to thwart the advance of postwar international communism, U.S. aid propped up the damaged economies of wartime allies and enemies alike. But now this aid is redounding to the detriment of U.S. trade balances.

These shifts in U.S. fortunes also are related to the worldwide acceleration and spread of the scientific, technological, and engineering enterprises. Scientific discovery, which involves theories about relationships found in nature, takes place in the context of a global community stretching from New York to New Delhi. Even when Einstein developed the insights that eventually led to production of the atomic bomb there was a substantial global scientific community, which had little difficulty understanding his formulas and their implications. Technology involves the application of scientific knowledge to solving practical problems. In the case of Einstein's discovery, the countless steps between abstract theory and a working bomb lay in the domain of technology and presented a barrier to constructing nuclear weapons for all but the most technologically sophisticated countries. And engineering involves the final nuts-and-bolts production steps that must be taken to transform scientific and technological discoveries into products.

Whereas scientific theories have floated rather freely through this international scientific community, technology and engineering have been much more proprietary, and the U.S. edge in these areas sustained its high-technology economic competitiveness for more than three decades after the war. More recently, however, the U.S. competitive position has been eroded by the rapid international diffusion of technological and engineering skills. Such knowledge now spreads rapidly across borders and is no longer dominated by one country. The economic resurgence of Japan, Germany, Great Britain, France, and other OECD countries has been one result of this more even distribution of science and technology capability worldwide. Even the newly industrializing countries are now capable of mounting concerted research efforts, by means of which they can keep pace with the more advanced countries. Thus, it is not surprising that major technological breakthroughs now take place in several different countries almost simultaneously. Big discoveries in the area of superconductivity, for example, were reported almost simultaneously in 1987 in the United States, Japan, Great Britain, Korea, and even China.[12] The circulation of ideas and experience through the interaction of scientists and engineers from many different countries by means of university training, exchange agreements, international meetings, and scientific publications has forged a worldwide scientific, technological, and engineering community; information moves through this community very quickly. In addition, the distinctions among science, technology, and engineering are increasingly blurred as laboratory findings are quickly translated into processes and products by public and private efforts to gain a competitive edge. In the case of superconductor discoveries, within days after basic new discoveries were made scientists in Japan and the United States were literally camped out in their laboratories attempting to be the first to turn discoveries into usable products.[13]

Taken together, economic and technological resurgence from wartime destruction helps explain the growing equality of industrial countries and the greater intensity of international trade competition. The parochial view from the United States is one of loss of foreign markets, a surge of new imports, unemployment, and a decline in economic fortunes (see Table 7-2); but a broader perspective on the situation would suggest the inevitability of adjusting to a more diverse and pluralistic system, no longer dominated by one country. Such a perspective, however, doesn't ease the pains or reduce the political difficulties of the major transition facing the United States.

The sharpening trade conflicts and related economic dislocation experienced by the United States also result from significant surplus capacity in important industries worldwide. In the absence of any superordinate international authority to regulate investment in new plants and equipment, individual nations and multinational corporations make capital investment decisions that may seem wise from a national perspective but may play a significant role in creating future, mutually destructive overproduction. Susan Strange has suggested several reasons that surplus capacity already bedevils international markets. On the supply side, required capital investment per unit of output has increased, thus shaping pressures to reduce idle capacity in existing facilities. In addition, a number of newly industrializing countries have sought to expand their domestic industries by entering markets once dominated by OECD countries. And on the demand side, a significant slowing of growth has been related to the dislocations of the two oil crises and associated factors.[14]

The global diffusion of science, technology, and engineering provides the underpinnings for the development of this surplus capacity. In the aftermath of the second world war, most of the world's research and development took place in the United States and in only a handful of other industrial countries. Discoveries often were closely held by research institutes and corporations. But it is now very difficult to

Table 7-2 U.S. merchandise trade balance (figures represent billions of dollars).

With:	1980	1982	1984	1986
Japan	−10.4	−17.0	−37.0	−54.4
Canada	−1.3	−9.3	−14.6	−13.3
W. Europe	20.4	6.8	−15.2	−28.4
OPEC	−38.2	−10.9	−13.1	−8.4
E. Europe	2.7	2.7	2.1	.5

SOURCE: *Economic Report of the President 1988.*

keep economically important discoveries proprietary for long unless they deal directly with highly secret defense matters. And experience has shown that some corporations will sell such technologies even in the face of obvious national security implications. The Japanese and Norwegian sale of quiet submarine propulsion technology to the Soviet Union in the mid 1980s is a case in point.[15]

While a rapid global diffusion of knowledge provides the underlying capabilities for countries to become competitive in key industrial sectors, nationalistic development policies, including government subsidies to and protection for new industries, provide incentives. A time trap is being created by the competing ambitious expansion plans in the newly industrializing countries, which may be rational and politically expedient at present but are creating excess global capacity for the future. Thus, in addition to the existing mercantilist drives to export in many of the industrial countries, growth policies in certain less developed countries are creating significant surplus capacity in many of the world's key industries. Between 1978 and 1985, for example, exports from Mexico grew about 300 percent, and those from Brazil, China, Hong Kong, and Turkey grew more than 150 percent. United States exports grew about 50 percent during the same period.[16] Since there is no global political authority to apportion production capacity among the world's national units, each sovereign country seeking to protect its national interest—and the fortunes of its political leaders—expands production capacity in domestically important industries as rapidly as possible.

Surplus capacity and resulting structural dislocation is thus a general problem for countries at all levels of development. While many of the less developed countries are locked into their own grim surplus capacity struggle as they export raw materials into a relatively stagnant market, their newly industrializing counterparts are in a similar struggle to capture a larger share of saturated world markets for manufactured goods.

In the face of increased manufacturing capacity for consumer goods, demand for them has been slackening. Prior to 1970, the global economy perked along at real growth rates in excess of 3.5 percent per year. Over the last fifteen years, however, the average growth rate has been below 2 percent per year, and in the worst years the world economy failed to grow at all. Several factors contribute to the slowdown. In the industrial countries, the approach of zero population growth is certainly a major factor in curbing increased consumption. ZPG is environmentally laudable but wreaks havoc with traditional growth patterns. In addition, however, the graying of Western Europe and the United States undoubtedly cuts into markets, since the elderly tend to be more conservative in spending habits and have less money to spend. Finally, although much has been made of problems of planned obsolescence and related shoddy workmanship, more intense worldwide competition seems to have produced better quality merchandise, which is more durable and needs fewer repairs. Thus, turnover of stocks of artifacts—automobiles, televisions, refrigerators, and so on—may be much slower than in the past.

Even consumption in the developing world, which should provide rapidly expanding markets, has lagged behind expectations. Although pent-up demand for manufactured goods in the developing countries is tremendous, it is not effective demand because of lack of purchasing power. Countries offering huge potential markets, such as Brazil and Mexico, are deeply mired in debt, and because of defaults in payments they are unlikely to receive infusions of new capital from outside sources in the near future.

Thus significant excess capacity now exists in a technologically more egalitarian world, and the dynamics of this time trap suggest that trade conflicts could become worse in the future. In the automobile industry it is predicted that overcapacity will be between six and fifteen million vehicles in the early 1990s. Detroit's domination of the world auto industry has been eroded by exports from Japan's nine auto makers, and these established Japanese companies are now in turn under attack by Hyundais from Korea, Yugos from Yugoslavia, and exports from Malaysia, Taiwan, and Thailand. The world steel industry faces a similar plight, with more than 150 million metric tons of capacity going unused. Almost all large developing countries have established a steel industry as a matter of national pride. Brazil, for example, produced 7.3 million metric tons of steel in 1976 and 17.3 million metric tons a decade later; most of the increase went into the export market. And it isn't easy to mothball old facilities. Quite apart from the politically sensitive unemployment problem, it is estimated that closing a typical mill costs a manufacturer nearly three hundred million dollars in worker payoffs and other expenses.[17]

High-technology industries are plagued with similar problems. Computer production has outrun expansion of the world's market. IBM, the traditional kingpin, has seen sales slump. New producers in the Far East, which used to sell mainly peripheral equipment, have been able to gear up for export of total systems at prices much lower than those of the established competition. This market saturation has been reflected in a major slump in the semiconductor industry and a bitter feud between the United States and Japan. Although the industry has recovered somewhat from the depths of the slump in 1985, when only about one half of plant capacity was being utilized, there is still little agreement between Japan and the United States on what constitutes fair pricing in the industry.[18]

▼ Industrial Policy or Nonpolicy?

Deterioration of the U.S. position in world trade and the related economic dislocation of the last decade have sparked heated political debates over the rules by which international commerce is conducted. Worried questions have also been raised about how and why the existing system in the United States has precipitated such a fall from preeminence, and what can be done to fix the situation and restore U.S. competitiveness. And the remarkable successes of Japan and the Asian NICs in capturing markets for manufactured goods have also raised questions about whether the United States should attempt to emulate their social and economic systems.

The trade rivalry between the United States and Japan is at the center of many imbalances in the world markets. Japan accounts for more than one third of the total U.S. trade deficit and a significant share of the rest comes from trade with the

Asian NICs. Part of the problem in U.S.–Japanese trade relations is rooted in the contrasting historical-ecological experience of the two countries that is reflected in contemporary institutions, beliefs, and behaviors.

Japanese population growth and industrial development have taken place in the face of acute natural resource scarcity. At present, more than 120 million Japanese are densely packed into a land area that is 80 percent mountainous. Japan has almost no oil and natural gas, and large quantities of food must be imported from other countries. Even fresh air is scarce in Japan: over the last two decades polluting industries have been exported to other countries as a matter of policy.[19] Thus, Japanese international politics and economics as well as domestic social policy have been shaped by pressing natural-resource concerns. With the Japanese it is an article of faith instilled from childhood that the Japanese economy must be based on exported manufactured goods that earn revenue with which vitally needed resources can be purchased.[20] Government subsidies to industries with future export potential exemplify the impact this thinking has on domestic policy. Internationally, Japanese foreign policy has hardly been one of principle; instead it has been pragmatically oriented toward the development and retention of markets for manufactured goods and the protection of raw material sources.

Economic growth and political development in the United States have taken place under very different circumstances. The westward expansion of the frontier served as a safety valve, keeping population density down, opening vast tracts of new farmland, and eventually yielding massive quantities of natural resources. This natural-resource treasury combined with a large and growing domestic market kept the United States mostly isolated from large-scale international trade competition. In an environment of resource abundance—quite the opposite of Japan's experience—different types of institutions, practices, values, and behavior patterns emerged. The American experience shaped an isolationist, "cowboy" mentality toward world affairs. Laissez faire economic policies evolved and were effective in encouraging the exploitation of a wide open, new frontier. Now that both the U.S. ecological situation and world markets have changed, however, the United States is experiencing a great deal of difficulty in shaping policies that could permit continued prosperity in a much more competitive world.

Given the different experiences and historical priorities of the United States and Japan, conflicts over trade practices and the nature of a level playing field seem inevitable. Japanese and American priorities, approaches, policies, and beliefs regarding industrial policy and fair trade are often at loggerheads. And the recent shift in economic fortunes, with Japan becoming the world's chief creditor and the United States becoming the chief debtor, has focused attention on these differences. Although there is a tendency in the United States to explain this sudden shift in economic fortunes with simple answers—such as the overly strong dollar of the mid 1980s—the persistence of the massive trade deficit in the face of devaluation suggests that more fundamental factors are at work.

The strength of the Japanese and more recently the Asian NICs' export performance comes from the mobilization of these societies for export-oriented economic growth. All the countries in which the most dramatic export growth has taken place over the last two decades are similarly barren of resources. They are densely populated and possess few of the natural resources essential for economic growth. In order to compensate, the Japanese, Koreans, Taiwanese, and others have organized export economies as quasi-military operations. From top to bottom, the priority of economic goals, indeed the importance of them for national survival, is understood and accepted.

The differences between the Japanese and American approaches to economic growth and international competitiveness are particularly apparent when the two social orders are compared. In Japan the emphasis in social life is on achievement and a collective investment in activities that will ensure a prosperous national future. There is a clear superordinate national goal: survival in a very competitive international economy. This national goal drives Japanese educational policy. Students spend long hours in the classroom, and the upwardly mobile often enroll in a second shift of private evening classes in order to prepare better for the competitive exams that determine their futures. The Japanese competitive emphasis is also reflected in the economy: the savings rate is extremely high and capital formation is rapid. In brief, the Japanese historical experience has shaped a whole society devoted to national economic survival through international trade in manufactured goods.[21]

Japanese social and economic practices stand in stark contrast to those in the United States. There have been no major external challenges to U.S. national survival in the last hundred years, and Americans lack clear-cut goals and competitive orientations. The historical atmosphere of resource abundance and laissez faire economic growth has created a liberal society devoted to individual rights rather than collective responsibilities and to immediate gratification rather than investments in the future. A decentralized educational system stresses adjustment, and competition is often discouraged. Furthermore, Americans spend a great deal of effort in relatively nonproductive maintenance activities. For example, by the year 2000 the United States will have nearly one million lawyers busily absorbing capital and manpower in a litigious society.[22] Japan has only a handful of lawyers, since to the Japanese way of thinking lawyers produce little that is of direct economic benefit.

There is also a significant difference in the way Japanese and Americans view the relationship between government and business. In Japan this relationship is nourished and is reflected in very explicit industrial policies that are developmental and expansionary. By contrast, in the United States this relationship is largely regulatory and restrictive and could best be depicted as industrial nonpolicy.[23] The Japanese industrial system is organized around the unquestioned need for export promotion to pay for the energy, food, and raw materials essential to Japanese prosperity. Production is designed for world markets, and government, banks, and industry closely cooperate in making investment decisions. Collaboration between government and firms, or even among firms, is generally harmonious given the well-understood international goals on which all agree.

Japanese industrial policy is thus forged in a collaborative relationship among government, banks, and industry, all seeking to move capital and manpower into promising areas. Firms cooperating with the government can expect to receive research and development funds as well as special treatment from the Ministry of Finance and the Ministry of International Trade and Industry (MITI). MITI is responsible for making judgments about the industrial sectors to be developed, and the Ministry of Finance works with banks to provide necessary capital. Although mistakes are occasionally made when some sectors are targeted for priority treatment, the Japanese are quite good at using their expertise and institutions to anticipate future problems.[24] Firms cooperate with MITI because of its power to influence allocation of bank credits, tax and revenue policies, import licenses, government subsidies, export allowances, and many other policy instruments.[25]

There are no analogous institutions or policies in the United States. In the pluralistic liberal society that has evolved in response to industrial affluence, government and business are officially kept at arm's length. There is little consensus on industrial policies, because no generally accepted national goals exist to facilitate such collaboration. Furthermore, heavy government spending on military research and development precludes a significant effort on the civilian side. This is not to say that government plays no role in industrial growth. The key differences between the role of government in the United States and Japan lie in motivation and clarity of purpose. The Japanese know what their long-term collective goals are, and they are motivated to achieve them. Policy makers in the United States are still mired in the corruption and muck of partisan politics and such politics precludes rational approaches to investment decisions.

In summary, it is unlikely that U.S. demands for a level playing field in international trade will have much of an impact on the future competitive situation. In Japan, Korea, Taiwan, and several other export-oriented countries, the care and feeding of export industries is a primary national concern. Thus, although these countries have expressed willingness to abolish border measures that are impeding international trade, they are unlikely to change important domestic policies, developed over decades, that facilitate exports and inhibit imports. Japanese politicians, for example, would be hounded from office if they voted to open the Japanese market to unrestricted imports of U.S. rice. And the Japanese are unlikely to dismantle MITI and all it stands for in order to please Americans. The tax incentives, export credits, low interest loans, and related measures that have made industrial policy so successful in these countries are unlikely to be abolished in the near future.

Although much has been said about increasing the competitiveness of the United States, of U.S. citizens and industry, little has been done to facilitate the social transformation required for this to happen. The feeble U.S. response to repeated massive trade deficits has been to debase the currency and attempt new protectionist legislation, which might allay some immediate problems but would make industries less competitive in the long run.[26] Critics of industrial policy continue to sing the praises of current nonpolicy and argue that industrial policies just cannot work in the United States. In some respects they can't. The political partisanship, regionalism, corruption, intrigue, and suspicion characteristic of contemporary U.S. politics make it extremely unlikely that politicians could pull together to make the necessary changes for meeting the competitive threat.[27] Thus the United States remains paralyzed in the face of these new challenges, leaders arguing over the desirability of industrial policies while avoiding the deeper question: what it would take to transform the present disparate liberal society to make competitive policies work.

▼ West–East Technology Transfer

Science, technology, and engineering are obviously important assets for national security. They are critical in providing an edge in the trade competition and conflict just discussed. And they are also central to the more direct, military aspects of security. The United States has relied on a strategy of technological innovation to increase the capabilities of limited manpower in hostilities in which adversaries have had superior numbers. Also, in recent years the United States has depended for security on maintaining significant high-technology lead time over its principal potential adversary, the Soviet Union. Given the acceleration of technological innovation and its more rapid diffusion throughout a global scientific community, however, concerns have been raised about how best to protect this national asset. Some people argue for clamping down on the flow of scientific information and trade with other countries in order to hoard U.S. expertise. Others argue that this is a self-defeating proposition because to do so would cut the U.S. scientific community off from international communication and force industry to cede high-technology markets to foreign competitors.[28]

Technology transfer refers to the various ways in which knowledge, machinery, and processes are transferred from one country to another. It is constantly taking place, in both intended and unintended ways. Intentional transfer takes place through exports, licensing arrangements, construction of turnkey production facilities or various forms of technical assistance. Unintentional transfer is a by-product of educational exchanges, articles in the scientific literature, scientific meetings, or even espionage. Technology transfer is a vehicle by which technologically less sophisticated countries can catch up and become competitive with their more developed counterparts. For competitive reasons, the technologically advanced countries want to take reasonable steps to protect their important assets, but opinion about how this should be done is divided. The chief debate in this area now focuses on the flow of scientific information and transfer of technology from Western nations to the technologically less sophisticated Soviet bloc.

There are deep differences, within both the Western alliance and the United States, over policies to restrict the flow of information and transfer of technology. In the current atmosphere of intense trade competition and conflict, the Soviet Union and other potential adversaries offer substantial markets for know-how, products, turnkey production facilities, and so on. The problem for sellers is one of coordination: defining a set of rules by which competitors can agree to regulate West–East scientific information flow, technology transfer, and trade.

Among Americans, there are several contending perspectives on scientific communication and technology transfer.[29] The most hard-line perspective, defended by vociferous opponents of contact and commerce, is that almost any transfer of technology by the United States or its allies increases the military capabilities of the Soviet bloc and should thus be prevented. Furthermore, they argue that by targeting certain technologies, the Soviet Union is able to avoid the cost of research and development that would be required to develop the same products domestically. At the other extreme are those who point out that scientific and technological progress can only take place in an open society free from undue restrictions on scientific communication within the country and across national boundaries. This position is associated with a more liberal approach to trade with the Soviet Union and other potential adversaries and is based on an understanding that trade restrictions hamper the ability of American industry to compete with foreign corporations.[30]

The debate in the United States parallels one within the Western alliance. In general, the United States has adopted a much harder line on West–East trade than most of the allies, who see normalizing trade relations as being to their economic advantage. In the early 1980s, for example, the United States fought a long, losing battle with Western European countries over their participation in the financing and construction of the Urengoi–Yamburg natural gas pipeline from Soviet Siberia to Western Europe. The U.S. argument was that the hard currency earnings from gas transported by the pipeline could be used to increase Soviet military strength. The Europeans saw the pipeline as an alternative to unreliable energy supplies from the Middle East, and the pipeline was completed only slightly behind schedule.[31]

The core issues in the dispute over information flow, technology transfer, and trade revolve around a definition of what flows and transfers are harmful to Western interests and the costs of preventing those transfers from happening. There is little disagreement that sensitive military technologies and related scientific information should be carefully protected. And there also is a fairly solid consensus that a wide array of goods and technologies with no apparent military value can be sold and transferred without risk. The debate focuses on dual-use technologies, which by their nature have both civilian and military applications. There is little agreement on where to draw the line in the dual-use areas. At one time or another, U.S. officials have: seized Apple computers bound for Eastern Europe on the grounds that they could be useful in military applications; berated the Japanese for permitting the export of medium-sized trucks to Vietnam, since they could be used as mobile missile launching platforms; and threatened to restrict the distribution of photos taken by privately owned earth satellites. The Austrians, by contrast, have had few qualms about transshipping all sorts of high-technology items to Eastern Europe.[32]

The United States attempts to control the domestic export of sensitive items by requiring export licenses for them. In recent years, about 40 percent of all exports of manufactured goods required such licenses. The unwieldy process of obtaining them often puts U.S. businesses at a competitive disadvantage with foreign firms that can promise more immediate delivery of similar products. The government also complicates matters by dividing recipient countries into various security groups. Products bound for one country might not be on the approved list for others, thus creating a lucrative opportunity for transshipment. The United States also attempts to force foreign purchasers of restricted U.S. technologies to apply for U.S. export license if they, in turn, export any form of the technology to a third country. This has been a major point of contention in efforts to enlist European countries and Japan in joint research efforts with the United States. Their fear is that jointly developed technologies would be subject to U.S. regulations and thus would not be profitable undertakings.[33]

The Western allies attempt to coordinate export policies through the Coordinating Committee for Multilateral Export Controls (COCOM), which consists of most NATO countries plus France and Japan. Although the activities of COCOM have received considerable attention in recent years, the organization is incapable of coping with the flood of controlled technologies and exports coming from member countries. Furthermore, not all members of COCOM are avid supporters of tight trade restrictions, and several key countries capable of transshipping sensitive equipment are not part of the organization.[34] Given the highly competitive nature of contemporary international trade, there are substantial incentives for circumventing existing understandings.

The ideological passions surrounding the political aspects of West–East technology transfer unfortunately obscure some of the more fundamental social science issues involved. Given the widespread diffusion of discovery and innovation in the global scientific community, it is both undesirable and impossible to restrict exports of much dual-use knowledge and equipment to industrial countries without dramatically handicapping American business. But technology transfer is also becoming more of a two-way street. Many discoveries are no longer made in the United States, and even those that are made there are quickly duplicated in other parts of the world. Even the Soviet Union has become a source of reverse technology transfer in certain specialized areas.[35]

These considerations have led distinguished panels to recommend new approaches to scientific information flow and technology transfer. In general, they have suggested that the U.S. government build strong fences around narrow areas—in other words, that it decide which information and which technologies are really vital and expend resources to protect them.[36] These areas to be protected should have direct military applications, potentially give the USSR a significant near-term military advantage, be in a rapidly developing area where time from basic science to application is short, and be solely in the hands of the United States or other friendly nations having controls as secure as ours.[37] Furthermore, given the economic competition aspects of the problem, the goal of U.S. policy should be to improve the multilateral control system by concentrating on the most critical technologies.[38]

Unfortunately, these recommendations have been ignored by right-wing politicians, and there have been increasingly broad but ill-conceived attempts to restrict trade and the flow of scientific information. Efforts have been made in the United States, for example, to keep foreign visitors from scientific meetings, university classrooms, and research laboratories. These actions are self-defeating because they lead to reciprocal measures by other countries. Furthermore, the ultimate effect of such actions would parallel that of possible protectionist legislation isolating American industry. In both cases, the United States would likely forfeit leadership by removing itself from international contacts and competition. Barriers to information dissemination also could have a chilling effect on scientific communication within the United States, thus handicapping American scientists and engineers. Finally, there is a very basic philosophical conflict in the idea of an open society broadly restricting the flow of information. To adopt stringent controls on such communication emulates the Soviet closed society model, which has not been remarkably successful in facilitating scientific progress.

▼ Managing Cutting-Edge Technologies

Rapid technological change has now become institutionalized in industrial societies. Increases in commonly measured living standards are one result; continuous structural change is another.[39] There are thus two different ways to view the impact of a worldwide acceleration of discovery and innovation. The first face of change is the obvious and much celebrated one that promises to raise global living standards, eliminate hunger, and prolong the life span. But the other face is less celebrated, because it is the more disruptive one that undermines established values and institutions. Just as the innovations of the industrial revolution destabilized and continue to destabilize traditional agricultural societies, contemporary cutting-edge technologies are beginning to challenge and transform values and institutions in industrialized countries.

Technological momentum has created two serious imbalances that impact both domestic governance and international relations. The first is an imbalance between the growing power of science and technology to reshape social institutions and values and the declining ability of liberal, pluralistic, politically divided societies to guide and regulate such activity and its impacts. An autonomous and uncontrolled technology is not necessarily in the long-term public interest, at least as viewed from a current value perspective.[40] A related imbalance exists between the pace of discovery in the physical and social sciences. The physical sciences have blossomed, but on the social and behavioral side there has been relative stagnation.[41] The technological societies about which critics frequently complain can be seen as inevitable products of rapid advances in the physical sciences that cannot be matched by social innovation and new political capabilities. In the United States, this imbalance is obvious in the rapid growth of intrusive high-technology methods of keeping social order that are moving into the void created by a national failure to instill social consciousness. Sophisticated lie detectors, complex surveillance devices, mandatory drug tests, and a host of other devices threaten traditional concepts of human integrity; they are increasingly employed as a substitute for advances in social and behavioral science as well as good governance.

The rapid diffusion of scientific and technological knowledge in a competitive and shrinking global village is creating similar international coordination and management questions unprecedented in international politics. Few established practices regulate the diffusion of these technologies among sovereign nations, and no existing institutions are strong enough to undertake such regulation.

In several cutting-edge areas of technological innovation, the new capabilities being developed force profound ethical and pressing regulatory questions. Three areas of activity are chosen here for closer analysis: nuclear science, biotechnology, and telecommunications. Innovations taking place in each of these three areas represent different kinds of challenges to the existing international regulatory order. Some projected capabilities are potentially so risky and destructive that an international prohibition on certain types of research and technology transfer might be appropriate. Various aspects of biological research leading to creation of new life forms that could potentially disrupt the global ecosystem are obviously candidates for close scrutiny. For other cutting-edge developments, the key question is one of preventing the transfer of existing know-how and equipment. Various types of innovations in nuclear technology fall into this category. Finally, a host of delicate questions about the cultural impact of certain kinds of developments must be considered. Various new capabilities in telecommunications raise these types of issues.

▼ Nuclear Proliferation

The management of nuclear technologies has been the prototype case in dealing with diffusion of potentially dangerous know-how and the related problems brought on by disruptive innovations. The development of nuclear technologies was accelerated by the second world war and desires of U.S. policy makers to create a weapon capable of ending the war with a minimum loss of American lives. Once the nuclear genie was let out of the bottle, however, it became very difficult to contain the technological and engineering knowledge necessary to make a weapon.

When Albert Einstein first spelled out the theories and equations demonstrating the feasibility of making an atomic bomb, these ideas were widely known in the scientific community. The difficulties for potential bomb builders, however, lay in the technologies and engineering necessary to move from theory to a functioning weapon that could be dropped from a plane. Containing the spread of weapons knowledge was difficult because very similar technologies could be used to generate electricity. Many of the scientists involved in making the first nuclear weapons even sought to demonstrate the value of their work through "atoms for peace" initiatives. The tricky question became how to permit some international diffusion of the know-how required to develop nuclear power while restricting the closely related technologies used to make weapons.

Electric power is produced by a nuclear reactor through a fission process analogous to that which occurs in a nuclear weapons explosion. In both cases enriched uranium or plutonium atoms fission, or split, when struck by neutrons. Each fissioning of an atom produces an average of 2.5 additional neutrons, gamma rays, other fission products, and heat. The neutrons thus freed can potentially strike other atoms and create a chain reaction that produces either heat to make steam or a powerful explosion. The main difference between a chain reaction in a nuclear reactor and that in a bomb is the speed with which it moves. A nuclear weapon is constructed to maximize the speed with which neutrons split atoms and the chain reaction leads to an explosion in a fraction of a second. The chain reaction in a nuclear reactor, in most cases, is carefully monitored: neutron-absorbing control rods regulate the rate at which heat is produced. When these cadmium or boron control rods are fully inserted into a reactor core, they absorb most of the neutrons and dampen the chain reaction. It is possible for reactors to malfunction, and those at Three Mile Island and Chernobyl were destroyed when chain reactions couldn't be controlled with regular procedures.

Aside from the continuing potential for environmental disasters caused by reactor meltdowns, the spread of nuclear technology raises crucial national security issues. The fuels used in nuclear reactors and the processes by which they are made are very much the same as those used to make bombs. Most reactors use enriched uranium 235 (U-235) in their cores, although breeder reactors use plutonium. Both materials can also be used in fabricating a nuclear weapon. Naturally occurring uranium is composed of two uranium isotopes. The most abundant one is U-238, which makes up 99.3 percent of the mixture. Only .7 percent of the mixture is U-235. U-238 fissions at a slow rate and is thus not very effective in conventional nuclear reactors and in nuclear weapons. U-235 fissions much more rapidly and is thus the preferred fuel. This naturally occurring uranium mixture must be enriched, to between 2 and 94 percent U-235, to be highly effective in either reactor or weapons applications. Most reactors currently in use require uranium containing only a few percent U-235. The next generation of reactors, however, will use highly enriched uranium or plutonium. An international security problem arises from the fact that the same enrichment processes can be used to make reactor fuel or bombs.[42]

There are four methods of increasing the percentage of U-235 in the naturally occurring uranium mixture: the gaseous diffusion, centrifuge, nozzle, and laser methods. Gaseous diffusion is the method traditionally used in the United States. Three huge, aging facilities—in Oak Ridge, Tennessee, Paduca, Kentucky, and Portsmouth, Ohio—service the nation's commercial reactors. The diffusion method of enriching uranium is very energy-intensive, and each of these facilities uses enough electricity to meet the needs of a modest-sized U.S. city. Centrifuge enrichment facilities are now in use in Europe. They have the advantage of requiring only about one tenth of the energy to do the same amount of work as the gaseous diffusion facilities, and they can be constructed on a much smaller scale. Centrifuge technologies thus make it easier to hide clandestine facilities. Nozzle enrichment is not a popular method. Laser enrichment is still an experimental technology, although the United States intends to replace its older facilities with laser plants in the future.

Aside from enriching uranium, weapons-grade materials can be obtained in two other ways using ostensibly peaceful reactor technologies. New generation breeder reactors create both power and the man-made element plutonium, an ideal material for use in nuclear weapons. Plutonium can also be recovered from waste by reprocessing spent nuclear fuel taken from conventional reactors. Countries possessing breeder

or reprocessing technology thus have the capability to manufacture the key ingre-
dient required for nuclear weapons. Although other complexities are involved in making weapons, there is general agreement that it is not very difficult for a dedicated
and well-financed group of terrorists to fabricate primitive nuclear weapons once
they obtain as little as fifteen kilograms of highly enriched uranium or four kilograms
of weapons-grade plutonium.

The United States dominated nuclear power development and sales from the
second world war until 1975, building 70 percent of the world's reactors and nailing
down two thirds of all export orders.[43] During that period, the United States and the
Soviet Union monopolized the enrichment business and were thus able to monitor
closely the spread of such technologies. But the first energy crisis focused attention
on alternatives to petroleum, and in the ensuing rush other countries—France, Germany, Britain, Canada, Sweden, and Japan—got involved in the development and
sale of reactors and components.

In the 1980s, partly because of reactor accidents, the high costs involved in
building reactors, and the decline in petroleum prices, the international market collapsed; reactor manufacturers were faced with extensive excess capacity.[44] Manufacturers have aggressively pushed sales to Third World countries. Thus, in addition
to the six countries that have exploded a nuclear device, three others—Israel, South
Africa, and Pakistan—are widely believed to have stockpiled some weapons, and
several more countries could go nuclear should they so choose. Because of technological advances and continuing threats of interruptions of petroleum supplies coming from the unstable Middle East, it is likely that the next century will see the continued spread of enrichment and reprocessing facilities and a great deal more plutonium
being transported from one country to another.

The institutional response to these technological challenges is the International
Atomic Energy Agency (IAEA), established under the auspices of the United
Nations in 1957. The main instrument it uses to control the spread of nuclear know-
how is the Nuclear Non-Proliferation Treaty (NPT), which was initialed by the
major powers in 1968. The NPT obliges signatory nations not to transfer weapons
or related technology to non-nuclear weapons states or to assist those states in developing weapons. The treaty requires signatory states to maintain records of their nuclear
materials, and the IAEA is charged with overseeing the agreement. A major problem with the treaty is that many nations involved in nuclear commerce are not
signatories, and the small IAEA inspection staff is overwhelmed by existing nuclear
commerce.

A less formal response to the proliferation problem has been the formation of
a group known as the London Suppliers Club. The club is composed of countries
that can supply weapons components not falling directly under the control of the
IAEA. It has been modestly effective in slowing the spread of nuclear weapons technologies, although Pakistan has apparently been able to buy much of the equipment
required for building bombs. Recently, however, the group has begun to focus on

limiting the spread of vehicles that deliver weapons, thereby tacitly admitting that their attempts to control the spread of bomb technologies have been less than a complete success.

Given the tentative nature of this nonproliferation regime, the question arises as to why only a handful of countries have demonstrated nuclear weapons capability. The many countries that could explode a nuclear weapon have chosen not to do so for a number of good reasons. The most obvious is that to explode a weapon is to risk reproach from the international community. Countries on the nuclear threshold have to calculate carefully the political and economic costs of going nuclear, since such a demonstration might lead to loss of foreign aid, loss of trade, or even diplomatic sanctions. A second factor in limiting the size of the nuclear club is the multibillion dollar cost of developing a weapons program. None of the technology involved in enrichment, reprocessing, or bomb construction comes cheaply, and to undertake such a program represents a major commitment of funds. Finally, for some countries it might be more useful to develop nuclear weapons capability without actually carrying out a demonstration explosion. For example, it is widely suspected that Israel has developed a nuclear weapons arsenal, but Israel has little to gain by confirming these suspicions through a weapons test.

▼ Regulating Biotechnologies

A second cutting-edge area of innovation capable of creating major international regulatory problems is in the biological sciences. The phrase *biotechnology revolution* refers to a wide range of discoveries promising greater human control over genetic evolution, diseases, and the physical environment. This revolution is unfolding in four areas. The most celebrated work is being done in the area of genetic engineering, where living organisms or parts of organisms are used to modify and improve plants and animals, to make new products, or to make microorganisms for specific uses. Using recombinant DNA technologies, scientists are developing capabilities to modify existing species of plant and animal life, to create new species, and eventually to alter the human body. Work is also moving forward in reproductive technologies. In vitro fertilization, sex selection of offspring, gene therapy, and a host of other innovations promise to give human beings much greater control over the reproduction process. A third area of activity focuses on transplants, creation of artificial organs, and development of various prosthetic devices. Finally, the large-scale development and use of antibiotics is changing the nature of the war against bacteria that threaten human and animal life.[45]

These advances in biotechnology promise to revolutionize health care through development of genetically engineered vaccines, new diagnostic techniques based on monoclonal antibodies, and therapies using materials cloned from peptides and proteins in the human body. Agriculture is expected to benefit by applying these techniques to farm animals as well as through enhanced production efficiency derived from the use of new growth hormones and similar products. In crop production, new microorganisms are being created to protect against insects, physical stress, and freeze injuries. A genetically redesigned bacterium, for example, promises to inhibit freezing in certain crops when it replaces the natural bacteria found on leaves. But the major benefits to agriculture are to come in the more distant future, as genetically engineered seeds begin to have an impact on world crop production. There are countless other applications for biotechnology, ranging from enhanced oil recovery to pollution control through the introduction of so-called designer organisms.[46] New reproduction technologies promise to be a boon to childless couples, and organ transplants offer the opportunity for improved human performance and lengthened life spans.

But this ongoing revolution in biotechnology poses ethical, economic, and political problems for leaders of industrial countries. The potential to change existing life forms and create new ones raises the question whether some types of activities should be limited. Experimentation with dangerous organisms must be closely monitored to prevent their release into the environment. The potential also exists for development of particularly destructive agents for biological warfare, and agreement is needed to limit that type of research.[47] But changes in existing life forms and creation of new ones is likely to be even more contentious. Should the human body be genetically altered to better perform some types of tasks, or should this type of activity be outlawed on a global scale? The new reproductive technologies open up possibilities for selecting the sex of offspring, but the new capability raises important social questions about the propriety or desirability of such selection. In many countries this would mean the growth, over time, of the percentage of males in the population and result in changes in values and institutions that are not yet understood. This revolution has already brought other legal and ethical questions into the open. Surrogate motherhood and the cryogenic preservation of embryos raise numerous complex questions of inheritance and rights to life. The booming business of organ transplantation raises ethical questions of access to organs. Should there be a market for organ transplants, or are there better ways to ration scarce resources?

These types of developments raise complex and delicate questions in countries with these new capabilities. But what are now domestic matters will soon spill over into international politics. It does little good to ban certain types of research in some countries if in others it continues to move rapidly ahead. Given the absence of universal ethical standards, research that may be ethically repugnant in one country could be enthusiastically embraced in another. And the resulting technologies and products can easily be transferred across national borders. Thus, the eventual worldwide spread of the biotechnological revolution makes international consensus on standards of conduct imperative.

Unlike the situation in nuclear technologies, no international organizations are charged with overseeing the development and transfer of biotechnologies. The OECD countries have consulted and come to some agreement on safeguards for recombinant DNA research.[48] But for the most part, policy regarding these technologies still remains in the hands of sovereign nations.

▼ Communication in the Global Village

The global revolution in telecommunications raises more immediate and somewhat more conventional regulatory problems. The development of powerful digital computers, communications satellites, and fiber-optic communications cables has created the capacity for instantaneous communication around the world. The typical current-generation communication satellite, for example, can handle up to thirty thousand phone calls and two television channels at one time.[49] This technology creates the potential for a constant worldwide flow of data, information, visual images, and the like. But it also creates the potential for challenges to national sovereignty, invasions of privacy, and squabbles over control of the communications spectrum.

The possible benefits of the global telecommunications revolution are enormous. Worldwide direct-dial phone service, teleconferencing, and satellite news reporting are among some of the earliest and most obvious uses of the new technologies. Less obvious but perhaps more important are their economic and political aspects. A worldwide stock and bond market is in the making as securities of major corporations are traded twenty-four hours a day on various markets. Capital flows freely by wire among the world's banking institutions. Multinational corporations can instantaneously transfer data from branches around the world to headquarters, and global marketing is fast becoming a reality. Much of the ordinary commerce within and among industrial nations is now carried via satellite, including major television networks and stations, commercial video services such as Home Box Office (HBO), bank transactions, telephone calls, and military communications. Satellites and computers aid in the instantaneous dissemination of photographs, blueprints, and data sets around the world. Doctors in remote regions can use satellite communications and computer programs to diagnose and treat illnesses that would never have been identified in the past. And the worldwide exchange of television, movies, and the like offers the opportunity for nations and their leaders to develop greater mutual understanding and theoretically to improve prospects for world peace.

Although their diffusion certainly isn't life-threatening, these technologies still create myriad contentious issues among nations and international regulatory and co-ordination problems. One of these is the issue of access to the telecommunications spectrum and the hardware used for satellite communications. The access issue is a peculiar kind of commons problem. The telecommunications spectrum is limited in the number of frequencies available for radio and television broadcasting. This problem has historically been handled through licensing procedures, which serve to limit the power of transmitters sharing the same frequencies in geographically contiguous countries. But the advent of communications satellites has sharpened the issue because of the very limited number of optimal frequencies available for "up and down" satellite communication. Most communications satellites are in geosynchronous orbit near the earth's equator at an altitude of about 22,300 miles. At this location and altitude their speed of rotation matches that of the earth below, thus keeping them in one spot in relation to the earth's surface. Given the constraints of existing technologies, satellites using the same frequencies should be spaced in orbit at least 4 degrees apart, thus limiting the number of parking places available. Furthermore, satellites tend to cluster in the parking spots that best facilitate communication between the United States and Europe and between the United States and Japan. Given that the United States and the Soviet Union alone plan to have nearly two hundred communications satellites of various kinds in orbit over the next decade, satellite traffic congestion is a major problem.

The institution in charge of overseeing the celestial parking lot is the International Telecommunications Union (ITU), a specialized agency of the United Nations. Satellite issues are debated in an ongoing Space WARC conference, where potential claimants for available space are allowed to make their case. Rights to equatorial parking places are now a major issue between industrialized and Third World countries. Third World countries see the satellite locations in geosynchronous orbit as a scarce commons resource that should be equitably shared by all nations. They argue that since the industrial countries have already put satellites in so many locations, most of the remaining slots should be reserved for future Third World use. The industrial countries have retorted that future innovations in communications technologies will expand the number of parking places and that the Third World concerns are premature.[50] But since these anticipated technologies will undoubtedly be expensive and originate in the industrialized countries, further aggravating the North–South split, these issues remain a continuing point of tension among the countries involved.

Another set of politically and economically contentious issues is being driven by developments in telematics, the area where telecommunications, electronic information, and data banks converge. Given the increase in capacity to transmit data and the almost negligible cost of doing so, governments, multinational corporations, research firms, and others are increasingly moving information around the world. Multinational corporations, for example, need information on marketing prospects, sales, supplies, inventories, and finances for branches and subsidiaries located in other countries. Governments and other organizations attempt to get demographic and survey information about relevant populations. U.S. public relations firms, for example, often have the best data available on demographics and attitudes in foreign countries and even get involved in managing political campaigns for foreign politicians.

There are many reasons that developments in telecommunications have given rise to Third World demands for a new international information order as well as domestic legislation regulating transborder information flow in more than sixty countries.[51] Control of information flow is increasingly seen as a sovereign right of nations, whether the information is coming into the country from a foreign direct broadcast satellite or is being exported from the country by a research firm. Protection against vulnerability, both economic and political, is another reason political leaders seek to restrict information flows. For example, attitudinal data are considered very sensitive in many countries, and their export is often tightly controlled. Economic nationalism is also an important force in regulating foreign data gathering and transmission activities, particularly of multinational corporations. And finally, some countries regulate data transfers on the ground that they are likely to violate individual rights to privacy.[52]

The communications and media revolutions have combined to create other novel issues regarding intellectual property. Regulating copyright of printed matter on a global scale has always been difficult, and as other media have been developed such problems have been compounded. There is the nagging question of regulating access to signals beamed from satellites, such as HBO movies. Entrepreneurs in less developed countries have been actively disseminating such movies without permission. Other more serious issues concern forms of intellectual property such as computer software or operating programs, compact disks, videotapes, and the like. International agreements to cover these types of questions are obviously needed, but currently there is little international consensus on many of the more exotic forms of intellectual property.[53]

▼ Notes

1. John Platt, "Social Traps," *American Psychologist* (August 1973); see also John Cross and Melvin Guyer, *Social Traps* (Ann Arbor: University of Michigan Press, 1980), chaps. 1–2.
2. Problems of coordination are discussed in Robert Goodin, *The Politics of Rational Man* (London: Wiley, 1976); commons tragedies are discussed in Garrett Hardin, "The Tragedy of the Commons," *Science* (December 13, 1968).
3. See Garrett Hardin and John Baden, eds., *Managing the Commons* (San Francisco: W. H. Freeman, 1977).
4. Arthur Stein, "The Hegemon's Dilemma: Great Britain, the United States and the International Economic Order," *International Organization* (Spring 1984).
5. See, for example, Richard Barnet and Ronald Muller, *Global Reach: The Power of the Multinational Corporations* (New York: Simon & Schuster, 1974); Raymond Vernon, *Sovereignty at Bay: The Multinational Spread of U.S. Enterprises* (New York: Basic Books, 1971); see also Raymond Vernon, "Sovereignty at Bay: Ten Years After," *International Organization* (Summer 1981).
6. See "Japan's Investment Binge in Southeast Asia," *Business Week* (November 3, 1986); "Japan's Embarrassment of Riches in South Africa," *Business Week* (February 1, 1988); "The Tidal Wave That's Sweeping International Finance," *Business Week* (July 13, 1987).

7. See the special report on trade and pricing policies in world agriculture in World Bank, *World Development Report 1986* (New York: Oxford University Press, 1986); Walter Mossberg and Ellen Hume, "U.S. Proposes Nations Cease Farm Subsidies," *The Wall Street Journal* (July 17, 1987); Stuart Auerbach, "Farm Subsidies Have Global Impact," *The Washington Post* (July 13, 1986).

8. Douglas Sease, "Trade Tangle: Taiwan's Export Boom Owes Much to American Firms," *The Wall Street Journal* (May 27, 1987).

9. See Robert Reich, "Beyond Free Trade," *Foreign Affairs* (Spring 1983); Gary Huffbauer and Joanna Erb, *Subsidies in International Trade* (Cambridge, Mass.: MIT Press, 1984).

10. An argument cogently made by Susan Strange, "The Persistent Myth of Lost Hegemony," *International Organization* (Autumn 1987), pp. 555–559.

11. Arthur Stein, op. cit., p. 384.

12. See Lewis Thomas, "Scientific Frontiers and National Frontiers: A Look Ahead," *Foreign Affairs* (Spring 1984), pp. 966–976.

13. John Burgess and Dale Russakoff, "Two Different Cadences in the Superconductor Race," *The Washington Post* (May 20, 1987). See also Lester Thurow, "A Weakness in Process Technology," *Science* (December 18, 1987).

14. Susan Strange, "The Management of Surplus Capacity: Or How Does Theory Stand up to Protectionism 1970s Style?" *International Organization* (Summer 1979), p. 308. See also Susan Strange and Roger Tooze, eds., *The International Politics of Surplus Capacity* (London: Butterworth, 1980); Peter Cowhey and Edward Long, "Testing Theories of Regime Change: Hegemonic Decline or Surplus Capacity?" *International Organization* (Spring 1983).

15. See "A Leak that Could Sink the U.S. Lead in Submarines," *Business Week* (May 18, 1987).

16. Data from "It's up to Germany and Japan to Ease the Trade Trauma: But Emerging Economies Could be the Wild Card," *Business Week* (May 4, 1987). See also "The Asian NICs and U.S. Trade," *World Financial Markets* (January 1987).

17. "Glutted Markets: A Global Overcapacity Hurts Many Industries," *The Wall Street Journal* (March 9, 1987).

18. "Is it Too Late to Save the U.S. Semiconductor Industry?" *Business Week* (August 18, 1986).

19. Kiyoshi Kojima, *Japan and a New World Economic Order* (Tokyo: Charles E. Tuttle, 1977), chap. 6.

20. Kenichi Ohmae, *Beyond National Borders: Reflections on Japan and the World* (Homewood, Ill.: Dow Jones–Irwin, 1987), pp. 2–3.

21. For an overview of the competitive nature of Japanese society and the evolution of Japanese industrial policy, see Chalmers Johnson, *MITI and the Japanese Miracle: The Growth of Industrial Policy* (Stanford, Calif.: Stanford University Press, 1982).

22. See Jethro Lieberman, *The Litigious Society* (New York: Basic Books, 1981).

23. John Zysman and Laura Tyson, *American Industry in International Competition* (Ithaca, N.Y.: Cornell University Press, 1983), p. 17.

24. Ibid., chaps. 1–2.

25. See Jack Baranson, *The Japanese Challenge to U.S. Industry* (Lexington, Mass.: Lexington Books, 1981), chap. 2.

26. John Zysman and Laura Tyson, op. cit., pp. 422–427.

27. For a discussion of industrial policies in the U.S. context see Lester Thurow, *The Zero Sum Solution* (New York: Simon & Schuster, 1985), chap. 9.

28. Short summaries of the extreme positions are found in "Should Export Controls Be a Political Weapon?" *Business Week* (September 12, 1983). See also Floyd Abrams, "The New Effort to Control Information," *The New York Times Magazine* (September 25, 1983).

29. An excellent overview of these positions is found in Mitchel Wallerstein, "Scientific Communication and National Security in 1984," *Science* (May 4, 1984).

30. See Roland Schmitt, "Export Controls: Balancing Technological Innovation and National Security," *Issues in Science and Technology* (Fall 1984); Office of Technology Assessment, *Technology and East–West Trade: An Update* (Washington, D.C.: U.S. Congress, 1983).

31. See Bruce Jentleson, *Pipeline Politics: The Complex Political Economy of East–West Energy Trade* (Ithaca, N.Y.: Cornell University Press, 1986); Anthony Blinken, *Ally Versus Ally: America, Europe and the Siberian Pipeline Crisis* (New York: Praeger, 1987).

32. See "High-Tech Leaks: The U.S. Gets Tough with Vienna," *Business Week* (November 26, 1984).

33. "Europeans Protest U.S. Export Controls," *Science* (May 11, 1984).

34. See Gary Bertsch, *East–West Strategic Trade: COCOM and the Atlantic Alliance* (Paris: Atlantic Institute, 1983); Bruce Parrott, *Trade, Technology, and Soviet–American Relations* (Bloomington: Indiana University Press, 1985).

35. For an overview of this reverse transfer see L. V. Rodionova, "The U.S.S.R. and East–West Transfer of Technology," paper presented at the joint Soviet–American Seminar on the Management of Technology (Laxenburg, Austria: November 1985).

36. This is the suggestion in the report of the panel on Scientific Communication and National Security (*The Corson Report*). See National Academy of Sciences, *Scientific Communication and National Security* (Washington, D.C.: National Academy Press, 1982).

37. Ibid., p. 49.

38. See National Academy of Sciences, *Balancing the National Interest: U.S. National Security, Export Controls and Global Economic Competition* (Washington, D.C.: National Academy Press, 1987).

39. Richard Cooper and Ann Hollick, "International Relations in a Technologically Advanced Future," in Anne Keatley, ed., *Technological Frontiers and Foreign Relations* (Washington, D.C.: National Academy Press, 1985).

40. See Langdon Winner, *Autonomous Technology* (Cambridge, Mass.: MIT Press, 1977).

41. See Dennis Pirages, "The Unbalanced Revolution," in Nicholas Steneck, ed., *Science and Society* (Ann Arbor: University of Michigan Press, 1975).

42. For technical details see Paul Ehrlich et al., *Ecoscience: Population, Resources, Environment* (San Francisco: W. H. Freeman, 1977), pp. 430ff. Also see George Quester, ed., *Nuclear Proliferation: Breaking the Chain*, special issue of *International Organization* (Winter 1981).

43. See Office of Technology Assessment, *Nuclear Proliferation and Safeguards* (Washington, D.C.: U.S. Congress, 1977); Energy Research and Development Administration, *U.S. Nuclear Power Export Activities* (Washington, D.C.: U.S. Department of Energy, 1976).

44. Christopher Flavin, *Reassessing Nuclear Power: The Fallout from Chernobyl* (Washington, D.C.: Worldwatch Institute, 1987), pp. 47ff.

45. See Office of Technology Assessment, *Commercial Biotechnology: An International Analysis* (Washington, D.C.: U.S. Congress, 1984), chap. 3.

46. Ralph Hardy and David Glass, "Our Investment: What Is at Stake?" *Issues in Science and Technology* (Spring 1985).

47. Jonathan Tucker, "Gene Wars," *Foreign Policy* (Winter 1984–85).

48. See "Europe Splits Over Gene Regulation," *Science* (October 2, 1987).

49. Wilson Dizard, *The Coming Information Age* (New York: Longman, 1985), chap. 3.

50. For a detailed analysis of these issues and negotiations see Milton Smith III, "Space WARC 1985—Legal Issues and Implications" (Montreal: McGill University master's thesis, 1984). See also Marvin Soroos, "The Commons in the Sky: The Radio Spectrum and Geosynchronous Orbit as Issues in Global Policy," *International Organization* (Summer 1982).

51. Rita O'Brien and G. K. Helleiner, "The Political Economy of Information in a Changing International Economic Order," *International Organization* (Autumn 1980).
52. See Joan Spero, "Information: The Policy Void," *Foreign Policy* (Fall 1982).
53. For a more complete analysis of problems of intellectual property rights, see Office of Technology Assessment, *Intellectual Property Rights in an Age of Electronics and Information* (Washington, D.C.: U.S. Congress, 1987).

▼ Suggested Reading

JACK BARANSON, *The Japanese Challenge to U.S. Industry* (Lexington, Mass.: Lexington Books, 1981).

GARY BERTSCH, *East–West Strategic Trade: COCOM and the Atlantic Alliance* (Paris: The Atlantic Institute, 1983).

GARY BERTSCH AND JOHN MCINTYRE, eds., *National Security and Technology Transfer: The Strategic Dimensions of East–West Trade* (Boulder, Colo.: Westview Press, 1983).

ANTHONY BLINKEN, *Ally Versus Ally: America, Europe and the Siberian Pipeline Crisis* (New York: Praeger, 1987).

STEPHEN COHEN, *Uneasy Partnership: Competition and Conflict in U.S.-Japanese Trade Relations* (Cambridge, Mass.: Ballinger, 1985).

FREDERIC DEYO, *The Political Economy of the New Asian Industrialism* (Ithaca, N.Y.: Cornell University Press, 1987).

WILSON DIZARD, *The Coming Information Age* (New York: Longman, 1985).

ELLEN FROST, *For Richer, For Poorer: The New U.S.-Japan Relationship* (New York: Council on Foreign Relations, 1987).

ROBERT GILPIN, *The Political Economy of International Relations* (Princeton, N.J.: Princeton University Press, 1987).

PHILIP HANSON, *Trade and Technology in Soviet-Western Relations* (New York: Columbia University Press, 1981).

ANNE KEATLEY, ed., *Technological Frontiers and Foreign Relations* (Washington: National Academy Press, 1985).

NATIONAL ACADEMY OF SCIENCES, *Scientific Communication and National Security* (Washington: National Academy Press, 1982).

NATIONAL ACADEMY OF SCIENCES, *Balancing the National Interest: U.S. National Security, Export Controls and Global Economic Competition* (Washington: National Academy Press, 1987).

OFFICE OF TECHNOLOGY ASSESSMENT, *Commercial Biotechnology: An International Analysis* (Washington, D.C.: U.S. Congress, 1984).

OFFICE OF TECHNOLOGY ASSESSMENT, *Intellectual Property Rights in an Age of Electronics and Information* (Washington, D.C.: U.S. Congress, 1987).

KENICHI OHMAE, *Beyond National Borders: Reflections on Japan and the World* (Homewood, Ill.: Dow Jones-Irwin, 1987).

BRUCE PARROTT, *Trade, Technology, and Soviet-American Relations* (Bloomington: Indiana University Press, 1985).

HUGH PATRICK, ed., *Japan's High Technology Industries* (Seattle: University of Washington Press, 1987).

RICHARD ROSECRANCE, *The Rise of the Trading State* (New York: Basic Books, 1986).

LEONARD SPECTOR, *Going Nuclear: The Spread of Nuclear Weapons 1986–1987* (Cambridge, Mass.: Ballinger, 1987).

SUSAN STRANGE AND ROGER TOOZE, eds., *The International Politics of Surplus Capacity* (London: Butterworth, 1980).

8

Defending
Future
Generations

Previous chapters have laid out a theoretical framework for anticipating changes in the global system and the development of related global issues. The analysis has been anticipatory and descriptive. Numerous problems, anomalies, and dilemmas characteristic of a nascent post-industrial international system have been identified. The forces of ecological and technological change have been isolated as driving potentially revolutionary shifts in the structure of relations among nations and in the values that guide human behavior. The advanced stages of the industrial revolution have produced one of the golden periods of human history, and the liberal values and institutions that now guide human behavior embody some of the noblest aspirations of the human race. But the push of large-scale ecological changes and the pull of a variety of new technologies now challenge these cherished institutions and values.

The focus and purpose of this chapter are very different. It concentrates more directly on the future domestic and international consequences of the changing U.S. role in this shifting international order. An assumption is made that policy is important and U.S. leaders can be more than mere pawns in a global game rigged by ecological and technological change. Sources of U.S. decline are identified and some policy choices available to the United States to ease the growing burden on future generations are outlined.

In the last decade of the twentieth century, the vast majority of the rapidly growing portion of the human race finds itself running much harder just to stay in place on the contemporary economic treadmill. In the Third World, rapid population growth, lagging demand for basic commodity exports, heavy indebtedness, and lagging foreign aid have cooled the engines of economic progress and increased unemployment and civil unrest. But even in much of the industrial world, and particularly in the United States, the last fifteen years have been a period of economic stagnation or decline in real individual income and undoubtedly more rapid decline in elusive quality-of-life indicators. In only a few years the United States has seen a reasonably balanced trade situation deteriorate into a series of massive deficits. In the same period, the national debt has doubled, and the U.S. has moved from being the world's largest creditor country to being its largest net debtor.

This period of shattered expectations and uncomfortable adjustment to new economic realities has been characterized by political uncertainty and a search for principles and values to replace those no longer seeming to provide satisfactory guidance. In some parts of the less developed world, guidance and certainty have been found by embracing Islamic fundamentalism and the worldview it suggests. In the United States, a slightly different but functionally equivalent type of religious, social, and political fundamentalism has surged; it gives comfort to those who wish to retreat from the complexities of choice in a rapidly changing world.

One source of the insecurity in contemporary domestic and international politics lies in the race between ecological discontinuities and technological change, on the one hand, and the ability to develop new institutions to cope with them, on the other. While enhanced ecological, structural, and policy interdependence has resulted from increased contacts among nations, the capacity to anticipate and collectively deal with resulting frictions has not noticeably increased. It is particularly unfortunate in this regard that the United Nations, the one great institutional hope for peace that developed out of World War II, has been weakened substantially by the failure of the United States to use and support it. A second major source of insecurity and instability is the vacuum created by the diminished political, economic, and moral leadership of the United States in world affairs. Not only has its role as the global hegemonic power substantially deteriorated, but the quality of U.S. leadership and its ability to formulate rational and coherent policies have seemingly diminished apace. At a time when principled and visionary international leadership have been most needed, the United States has produced a series of presidents having little background in or understanding of international affairs.

The nationalistic politics of muddling through, characteristic of the resource-rich industrial economic expansion, will be inadequate to cope with the global issues of the twenty-first century. The industrial politics of conquest and expansion must be replaced by a post-industrial, global, anticipatory, inclusionist politics aimed at designing an equitable and sustainable future for the whole planet. The present generation is the first to develop an understanding of the causes of human misery and its planetary dimensions. It is also the first to be capable of directing scientific, technological, and socioeconomic efforts to eliminate misery and redress grievances. But to do so requires developing the political will and institutional capacity to design preferred futures and transition strategies to create them. Without such visions and the will to create them, relations among nations will continue to be dominated by the political economy of nationalism and the plethora of problems outlined earlier.

The declining economic and political fortunes of the United States are not only of interest to Americans but should be matters of concern to numerous client states. In an increasingly interdependent world, U.S. markets have become essential to the economic health of many other countries. The national interests of the United States and those of other countries are now intertwined in many ways. The most obvious is that the United States, the home of 250 million consumers, is still the most technologically sophisticated and economically important nation in the world. Although only 5 percent of the world's population lives in the United States, these people account for nearly 30 percent of the world's economic production and a very large share of the world's technological innovation. In the increasingly interdependent global economy, the decline of such a major economic force has to be viewed with trepidation by all key players. Should the economy of the United States collapse, it likely would take the global economy along with it.

The threat of an economic decline in the United States is of greatest immediate concern to numerous countries that have become highly dependent upon U.S. markets for their exports. Most of these countries are undifferentiated one-crop economies exporting most of their basic commodities to the United States. Table 8-1 lists the most vulnerable of them. But the health of the American economy is also important to several countries that seem much less vulnerable but are still dependent on U.S. markets. The reverse perspective on problems of U.S. trade with Japan, for example, is the potential Japanese net loss of a U.S. market worth fifty billion dollars annually.

Table 8-1 Highly dependent client states (figures represent the percentage of total country exports sent to the United States).

	1985	1986
Dominican Republic	83	83
Haiti	82	82
Congo	81	54
Canada	75	75
Mexico	67	60
Ecuador	64	54
Trinidad	62	63
W. Samoa	61	62
Barbados	53	53
Honduras	49	48
Belize	49	58
Costa Rica	46	40
Angola	45	45
Uruguay	42	15
Venezuela	41	41
Antilles	41	41
Guyana	41	22
Yemen	41	41
Japan	38	38

SOURCE: *International Monetary Fund, Directions of Trade Yearbook 1986, 1987.*

Similarly, many of the newly industrializing countries—Korea, Singapore, Taiwan, and others—depend on the United States for their own economic welfare. A U.S. economic collapse, or even a resurgence of protectionism, could significantly damage these fragile economies.

The interdependent global community must also be concerned about the fate of the dollar. It has been and remains the de facto world currency. OPEC oil sales are denominated in dollars. Many countries in the less developed world peg the value of their currencies to the dollar. Even the Japanese invest heavily in U.S. treasury bills, bonds, stocks, and other dollar-denominated assets.

Militarily, much of the West depends on U.S. leadership. Without very substantial U.S. military expenditures, Japan, Western Europe, and other protected nations would have to dramatically reassess defense expenditures.[1] Thus in a political, military, social, and economic sense, the health of the United States is integral to the welfare of the international system. The United States inherited a very important hegemonic role by default after World War II, and there are presently no obvious successors for the job.

▼ Mortgaging the American Future

The United States now faces a triple deficit threat of concern to Americans as well as their trading partners. The first and most obvious of these is a persistent and massive trade imbalance. Figure 8-1 details the rapid reversal of U.S. economic fortunes. Between 1980 and 1986 the U.S. current account balance plummeted from a small surplus to a whopping deficit in excess of $160 billion. While total exports stagnated, in constant dollars, during this period, imports nearly doubled. The overvalued dollar, a side effect of the Carter-Reagan war against inflation, undoubtedly played a large role in the loss of export markets, with total exports remaining nearly flat from 1980 through 1988. Imports have been adversely affected by America's growing demand for foreign oil, although big increases in automotive imports have been responsible for a considerable portion of the deficit.

Looking to the future, it does not appear that feeble U.S. responses, such as debasing the value of the dollar, will quickly turn the situation around. The weaker dollar has had some effects on the trade deficit, but other countries play the devaluation game at the same time. Thus, many U.S. industries have actually lost ground because the countries in which their competitors are located have been able to devalue their currencies at least as fast as the United States.[2] Even the much touted trade edge in high-technology products, the pride of U.S. industry, went into deficit for the first time in 1986 and is unlikely to recover substantially in the near future.[3]

The impact of this international pressure on U.S. industry and American workers has been considerable. Many manufacturing industries have closed their U.S. plants and moved production abroad, where labor is cheaper. Others, including the automobile industry, have been forced to cut production significantly in the face of foreign imports. Taken together, these moves have decimated high-paying blue-collar employment in the United States. While overall unemployment figures are now manageable, there has been a general shift of workers from high-paying factory jobs to lower-paying service jobs and a general decline in salaries. After adjusting for inflation, American workers averaged annual pay increases of 2.5 percent in the 1950s and 1.7 percent in the 1960s. But there have been no significant real salary increases since

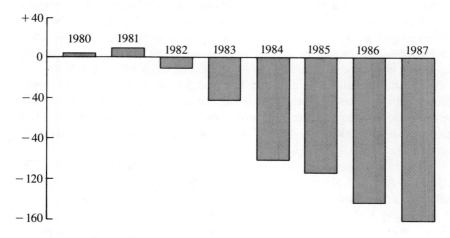

Figure 8-1 U.S. current account balance, in billions of dollars.
Source: *Economic Report of the President 1988.*

1970. Furthermore, since 1973, inflation-adjusted average weekly earnings have fallen more than 10 percent and median household income has dropped considerably. The average earnings in 1984 dollars for a thirty-year-old man were $11,924 in 1949. By 1973 this figure had jumped to $23,580. But by 1983 it had dropped to only $17,520.[4]

This erosion of American living standards may well be an inevitable adjustment to new competitive realities in the absence of a major transformation of American competitiveness. Economist John Culbertson has argued that unregulated foreign trade "throws peoples and nations with widely different wage levels and standards of living . . . into a destructive competition with one another."[5] He claims that the drain of U.S. jobs to the Third World is inevitable given traditional laissez faire approaches advocated by American free traders. He warns, furthermore, that once such a decline begins, it is difficult to support the social overhead required to regain a competitive edge, since a low income society cannot support the organizational structures and overhead requirements of advanced education and technology.[6] The record of the last fifteen years seems to support many of Culberton's contentions. Given the increasing international mobility of capital and technology, manufacturing industries have incentives to migrate to countries with cheap raw materials and labor supplies, thus continually draining jobs away from high-wage American industries.

It is frequently argued that these contemporary adjustments are healthy ones and that as manufacturing jobs are lost, new ones in the service sector are created. Unfortunately manufacturing matters; most of the new service jobs created are low-paying and produce little of consequence for international commerce. The largest increase in new jobs between 1984 and 1995 is expected to be in the service sector, with greatest demand for cashiers (556 thousand), nurses (452 thousand), janitors and cleaners (443 thousand), truck drivers (428 thousand), and waiters and waitresses (424 thousand).[7] They also don't have tight linkages with or multiplier effects on other sectors, as manufacturing does.[8] Even in the service sector, the United States is losing its edge as overseas competitors are now taking business away from U.S. companies.[9]

Pseudo-prosperity was maintained in the Reagan years through growth of a second deficit: massive borrowing from future U.S. generations as well as from other countries. Figure 8-2 details the growth of the federal government debt over the last decade. In the 1970s the federal government ran small and manageable deficits ranging from $3 to $59 billion annually. In the 1980s, however, the supply-side economic dogma of the Reagan administration led to massive federal deficits that peaked at $221 billion in 1986. When the Reagan administration took office in 1980, the accumulated federal deficit was about $900 billion. By the end of his second term in 1988, it had more than tripled to $2.6 trillion.

The debt increase has obviously been driven by political considerations. The $221-billion deficit in 1986 represented a consumption and employment subsidy of $900 for each U.S. citizen and a $3600 subsidy for a typical family of four. Another way of looking at the deficit is that such subsidies represent the taxes that would have to be collected to finance government services if the budget were balanced. The increase in the federal deficit has been politically palatable because it has taken place at the expense of future generations, and they have no way of influencing contemporary

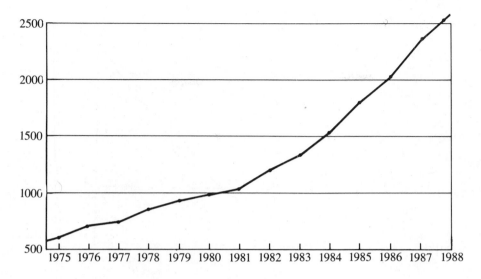

Figure 8-2 Total U.S. federal deficit, in billions of dollars.
SOURCE: *Economic Report of the President 1988.*

politics. This rapidly growing deficit means that, for the first time in U.S. history, future generations will probably experience lower wages, higher taxes, and declining standards of living.

The rising deficit and related changes in patterns of government obligation will severely restrict possibilities for a positive government role in fostering future investment. For example, just the cost of servicing the federal debt rose from constituting 8 percent of the budget in 1979 to 14 percent in 1988; this percentage will continue to rise along with unbalanced budgets or higher interest rates. In addition, in 1988 defense spending accounted for 30 percent of government expenditures, social security 22 percent, income security 12 percent, and medicare 7 percent.[10] These items taken together account for 85 percent of the federal budget, and much of what is left goes into various other direct consumption activities. Education, nondefense research and development, and a host of other activities necessary to restore U.S. competitiveness are virtually ignored in the federal budget.

Other obligations are also being incurred to cushion the decline in U.S. wages. At the federal level, future projected budget deficits look much worse if the short-term surplus being contributed by the social security system is removed from consideration, since it is earmarked for retirement benefits in the 1990s.[11] Various types of government loans are also being used creatively to keep them from showing up in the federal budget. In 1986 the government was responsible for more than $26 billion in delinquent loans to farmers, students, and various other domestic interests.[12] Debt is also rapidly growing in the private sector (see Figure 8-3). Household debt jumped from 24 percent of disposable income in 1976 to more than 31 percent in 1987. Personal bankruptcies grew from two hundred thousand to four hundred thousand annually during the same period. Even corporations have gone deeply into hock,

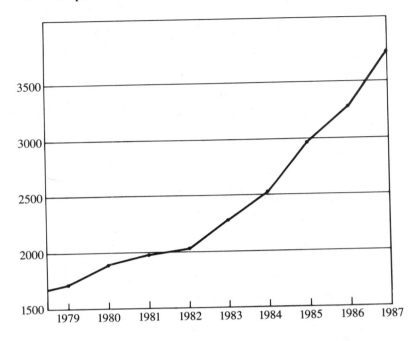

Figure 8-3 U.S. consumer and mortgage credit, in billions of dollars.
SOURCE: *Economic Report of the President 1988.*

with corporate debt now nearly 120 percent of corporate net worth.[13] In normal times the growth of such debts at least would be cause for concern. Given dimmer prospects for future economic growth in the United States, such mounting debts raise the specter of widespread bankruptcy.

The third deficit of concern is the meteoric plunge of the United States from its status as the world's number one net creditor into the cellar of the global poorhouse. The growing interdependence of the industrial economies has had a significant effect on the traditionally isolationist United States. In 1965, foreign trade (imports and exports) amounted to only 10 percent of gross national product. By 1985, foreign trade had doubled to 20 percent of GNP.[14] As recently as 1981, the United States was the world's largest net creditor, holding $720 billion worth of assets abroad while foreigners held only $579 billion worth of assets in the United States—yielding net assets of $141 billion. In only four years, investment flowing in both directions grew substantially, but persisting massive trade deficits drove the United States into the red (see Table 8-2). In 1986 U.S.-owned assets abroad totaled $1.1 trillion but foreign holdings in the U.S. had grown to more than $1.3 trillion.[15] The U.S. position has continued to worsen, and it is estimated that net U.S. international indebtedness could reach $1 trillion shortly after 1990.[16]

Given the huge U.S. trade and government spending deficits, an influx of foreign capital has been essential to maintaining economic viability. The persisting trade deficits have led to a much devalued dollar, which makes American assets very

Table 8-2 U.S. and foreign assets, in billions of dollars (small errors due to rounding).

	U.S. assets abroad	Foreign in U.S.	Net U.S. position
1979	511	416	95
1980	607	501	106
1981	720	579	141
1982	825	688	137
1983	874	784	90
1984	896	893	4
1985	949	1061	−112
1986	1068	1332	−264

SOURCE: *Economic Report of the President 1988.*

attractive to foreign buyers. Stocks, bonds, treasury bills, real estate, and even corporations are being snapped up at a rapid rate by foreign investors. The public and private sectors have become addicted to this steady flow of capital from abroad. In the mid 1980s, about 30 percent of the funds required to meet public and private borrowing needs originated abroad.[17] In addition, more than 15 percent of publicly held treasury securities are now owned by foreigners.[18] In sum, in the late 1980s foreigners held more than $1.5 trillion in assets in the United States.[19]

This foreign interest in the American economy has been an important factor in maintaining domestic stability; but it also now poses threats to U.S. economic and political independence. The addiction to foreign capital forces unpleasant economic policies on the U.S. public. To keep foreign capital interested in financing U.S. debt, interest rates cannot drop below a certain level. And as more stocks, bonds, treasury notes, and the like are owned by foreigners, the domestic economy becomes hostage to foreign interests. It is not beyond the realm of possibility, for example, that at some future date U.S. assets could be manipulated by other countries to express displeasure with U.S. foreign policy.[20]

▼ Sources of U.S. Decline

The rapid shift of the United States from international creditor to debtor status has resulted from three separate, related trends. First, the United States over time has become much more tightly integrated into a competitive world economic system. Second, as a natural response to decades of technological innovations, resource abundance, and rapid economic growth, social evolution in the United States has produced a permissive consumer society that promotes rights over responsibilities and immediate consumption at the expense of investment. Finally, precisely when U.S. industry has been forced to deal for the first time with substantial large-scale international competition, the techno-ecological sources of past growth and competitiveness have substantially eroded, raising the twin problems of adjustment to international competition while meeting the lofty social and economic expectations built up by decades of easy growth and prosperity.

Internationalization of the U.S. economy has been a slow and painful process, and the recent challenges from foreign competition are unprecedented in U.S. history. Buffered by oceans on the east and west and mostly friendly neighbors to the north and south, political leaders in the United States historically followed isolationist policies. Although two world wars and several peacekeeping operations involved the U.S. military in adventures abroad, it wasn't until the first oil crisis that Americans were forced to confront growing resource vulnerabilities and sharpened economic competition. Since then, the domestic economy has become increasingly dependent on foreign sources of energy and other resources as well as on foreign markets. In the mid 1980s, for example, imported oil accounted for more than one third of domestic consumption, and merchandise exports were equal to more than 10 percent of gross national product.

The permissive liberal heritage of past abundance is the second component of the U.S. problem. The United States is handicapped in this new and more intense international competition by the ease with which technological innovation and resource abundance fostered past economic growth. While Japan, Taiwan, Korea, and a host of other countries were learning to compete in resource-deficient settings, economic growth took place in an atmosphere of resource abundance in the United States. One result of this affluence and lack of challenges was the emergence of a laissez faire approach to growth and competition. Thus, U.S. industry is not now managed or motivated to be competitive with the Japanese, Koreans, and others who are much better organized. It is an unanswered question whether Americans are willing to undergo a traumatic change in values, life-styles, and industrial organization in order to compete. Currently, investment activity is overshadowed by immediate consumption, planning for the future is ignored in favor of short-term profits and perspectives, rights have priority over responsibilities, and an emphasis on pure equality overshadows equality of opportunity.

Apologists for the performance of this permissive, consumption-oriented society point out that the U.S. economy grew at a healthy rate in the late 1980s, but they ignore the quality and directions of this growth. Economists frequently praise the American transition from a goods-producing to a service society. But many of the currently growing services hardly bolster future economic growth or competitiveness. One major U.S. growth industry is litigation. Americans are increasingly hauling one another into court, and the United States has become known as the litigious society.[21] By the mid 1990s more than one million lawyers will be busily creating additions to GNP in the court system. U.S. courts and prisons have expanded rapidly in response to both a new legalism and a rise in crime. Advertising is another rapidly growing field that now makes a substantial contribution of dubious long-term benefit to GNP. While this explosion of consumption and maintenance activity has been taking place, investment in the collective future has lagged. Real investment in civilian research and development has remained flat, the educational system has deteriorated, and capital formation has stagnated. If U.S. leaders are really serious about making the United States more competitive, they will have to begin quickly reversing the habits of affluence that have led to these peculiar national priorities.

The third source of the U.S. predicament is the disappearance of industrial production advantages as a result of changing techno-ecological realities. These new realities are manifest in increasing shortages of and higher prices for the resources required for industrial growth, the graying of the American labor force, diversion of capital from investment to consumption, and a slowdown in civilian research and development.

The United States originally was blessed with an abundant supply of land and resources. Nature endowed the country with a huge treasury of fossil fuels, including petroleum reserves as large as those of contemporary Saudi Arabia. Iron, copper, and a host of other industrial minerals were also readily available to facilitate resource-intensive industrial growth. But more than a century of large-scale exploitation has eroded this natural resource endowment, and the cheap domestic resources that gave American industry an edge in competition with Japanese and European manufacturers no longer exist. It is premature to declare that the United States is rapidly running out of resources, but the competitive posture of American manufacturing is hurt by higher prices that must be paid for domestic raw materials. There is now general agreement among experts that most of the cheap domestic oil and gas in the United States has already been discovered. Over the last decade the percentage of crude oil imported from abroad has varied between one third and one half of domestic consumption on a monthly basis, giving added testimony to the disappearance of cheap domestic reserves. The mining industry has experienced diminishing returns from domestic mines, and much of the industry has migrated to less developed countries where ore bodies are richer and labor costs lower.[22]

Rapid progress in science, technology, and engineering have historically been responsible for keeping U.S. industry ahead of foreign competition and especially competitive in high-technology industries. But a decline in educational excellence combined with major efforts made by other countries have blunted any U.S. advantage. As mentioned in Chapter 7, internationally the postwar dominance of the United States in research and development has evolved into a more balanced situation in which numerous industrial countries make substantial contributions to R and D. In fact, in recent years nearly one half of all U.S. patents have been granted to residents of other countries.[23] More important, however, are the development of engineering skills in other countries and the rapid diffusion of technological innovation around the world. The competitive benefits of science and technology are no longer totally captured by the innovating countries but are now shared with emulating countries in which engineers are geared up to translate discoveries into products. Traditionally, the United States has maintained a healthy trade balance in high-technology products, but in recent years even this trade category has fallen into deficit.[24]

Changes in U.S. domestic priorities have also been responsible for reducing the contribution of science and technology to U.S. competitiveness. Tighter budget constraints have reduced capital available for civilian research and development and, as a percentage of GNP, civilian R and D has lagged significantly behind that of competitor nations.[25] Furthermore, heavy defense spending, including large sums devoted to "Star Wars" research, have siphoned off talent from other kinds of research that could increase U.S. competitiveness in merchandise trade. Finally, although precise national measures are hard to come by, there are indications that the liberal U.S. educational system, often emphasizing remedial educational programs, is no longer producing a large pool of talented, dedicated, and innovative scientific personnel.[26] Thus, because of both domestic permissiveness and increased international competition, the former U.S. competitive edge in high-technology products is rapidly dulling.

Finally, one other source of American growth in productivity has been readily available, cheap capital, but domestic investment capital is also now hard to come by. Savings rates in the United States are among the lowest in the industrial world, and interest rates that reached crippling heights in the early 1980s are still above those of competitor countries. High real interest rates discourage investment and innovation.[27] Furthermore, as the United States has become the world's largest debtor, an increasing portion of capital investment is controlled by foreign sources, meaning that many crucial investment decisions are now beyond U.S. control.

To sum up, when the traditional ecological and technological sources of American industry's edge in international competition are examined, negative trends show up in each case. Some of these negative factors—such as the diminishing natural-resource base or the growth of foreign technology and engineering—cannot be altered. Others, however—such as the lagging quality of the labor force and the educational system or the paucity of investment in civilian research and development—can be reversed by resolute action. The increasing and irreversible integration of the United States into the world economy forces tough quality-of-life decisions. Some have suggested that the only way to preserve the cherished liberal values of the advanced stages of the industrial revolution and to avoid being pulled down to the economic levels of the less developed countries is through strong protectionist policies.[28] Others, however, caution that these kinds of policies would lead U.S. industry into an increasing inability to compete with foreign manufacturers, a one-way street that would eventually have a dramatic negative impact on the U.S. consumer.[29] Whatever the remedies, it is clear that these problems must be addressed soon or the United States will continue its slide from the ranks of the superpowers because of paralyzed leadership and inadequate policies.

▼ Building Sustainable Prosperity

It is clear that many of the consequences of the transition from industrial expansionism to post-industrial limits and intensified international competition have so far been avoided in the United States by mounting up debts to be paid by future generations. Some of these debts are obvious, such as the triple deficit that will be passed on to the next generation of Americans. Others are less obvious, such as the steady decay of the physical environment from the demands of population growth and industrial development.[30] Although future generations don't vote, contemporary values dictate that the present generation has a responsibility for maintaining the quality of life for those who follow. In the past such concerns have been brushed off by technological optimists and exclusionists, who have argued that history shows each succeeding generation to be better off than its predecessors. But this comfortable assumption seems no longer valid when dealing with intergenerational equity questions during this period of transformation.

There are no easy answers to the problems and dilemmas facing the United States, and no laundry list of quick fixes can be laid out here. The typical response to many of these problems is to appoint yet another learned commission, which studies the issues and comes up with dozens of suggestions for bettering the human condition. The resulting reports are read, sometimes debated, and then are left to gather dust on library shelves because they lack any linkage with ongoing political processes.

This should not be taken to mean that deriving prescriptions from inclusionist analysis is useless, but rather that weak political institutions must be understood to be a major part of the problem. Assuming a willingness on the part of future governments to attack these deficits and problems, an inclusionist perspective suggests new directions for the United States, for the other industrial countries, for the developing world, and even for collective international action. Some changes, such as a major shift in U.S. definitions of national security, are so obvious that they should be immediately implemented. Others are more difficult because they fly directly in the face of established ways of doing things. But problem solving requires strong and intelligent political leadership. The commercial politics that have accompanied permissive liberalism in the United States have yet to produce leaders who have the knowledge, vision, and understanding to deal with these problems of transition in a resolute manner.

On the international level, much tighter interdependence and greater equality among nations makes industrial realism and power politics obsolete. Richard Sterling has suggested that these outmoded ideas be replaced by a macropolitical perspective in both theory and practice. This perspective stresses the unity of the international system and begins with "questions central to its global concerns: what is in the international interest? . . . What policies and institutions appear to benefit all men, and what appear to benefit some and not others?"[31] A macropolitical perspective represents an ecologically sound inclusionist viewpoint, which is a significant departure from the national interest–realism paradigm that evolved during the decades of industrial expansionism. It stresses the interdependence of all human beings and the complexity of emerging global issues and suggests that all nations must take these factors into consideration in foreign policy decision making.

Existing international institutions must be strengthened, and new ones built to deal with the problems raised by the many dimensions of complex interdependence. But this requires considerable surrender of sovereignty by the major powers. In the United States, the Reagan administration was particularly remiss in its decision to weaken and ignore the United Nations. A supporter of the UN as long as it was dominated by industrial country interests, the United States has since become embroiled unilaterally in several areas of the world while ignoring the necessity of multilateral support. Solving the global problems of the twenty-first century will require more multilateral action and less nationalistic posturing.

The United States bears particularly heavy responsibility for promoting the global welfare of future generations. The U.S. economy, by virtue of its size, will remain central to world prosperity, and future economic and political health or disease in the United States will be transmitted throughout the international system. For most of the postwar period the United States assumed global leadership by ascription. In the twenty-first century, U.S. leadership can be maintained only through good example and intelligent policies, keeping in mind the need for a macropolitical perspective in a world characterized by tighter interdependence. Recognition of ecological limits and the need to direct technological development into areas that will benefit the bulk of the human race will be priorities. The United States will never again be in a position to dominate a noncompetitive international system. Future hegemony will have to be earned through policies and actions perceived as being in the global interest.

Most important for the United States is to make the domestic changes necessary to adjust to a world of limited opportunities and more intense competition. The triple deficit that has resulted from unwillingness to deal with new global realities can only be dealt with permanently through major changes in social organization. Decades of nondecision have paralyzed U.S. political institutions, and a major overhaul will be required to succeed in a more frugal future. The choice before the American people is whether to continue to avoid confronting the causes of the deficits and drift into a more precarious and unstable future, or to begin a massive transformation of the permissive, consumer society that lies at the roots of U.S. competitive disadvantages. It may well be that the American people will choose to ignore these pressures and retain the laissez faire approaches in politics and economics that were more appropriate to the days of easy growth. But the deficit predicament at least forces a reassessment of U.S. goals and aspirations.

There is also a long agenda for the less developed countries. Most obvious in many of these countries is the absolute necessity to confront population crises. Particularly in the least developed world, little progress can be expected without a major effort to cut back birth rates. But this means confronting traditional forces and using educational systems and other socialization processes to change values. There are few nonintrusive ways of bringing about rapid value change. Furthermore, managing a revolution of rising expectations that in most cases cannot be met requires strong political leadership, which is lacking in many of the countries where it is needed.

In conclusion, numerous policy prescriptions can be derived from an inclusionist perspective, but there are no magic formulas that will permit far-reaching change without turmoil. Most such prescriptions, whether they apply to industrial or less developed countries, ultimately require changes in beliefs, habits, values, and behavior, not easily accomplished within the framework of existing stagnant political institutions and processes. If the harsh consequences of following the current dangerous trajectory in relations among nations are to be avoided, and future generations are to be left a decent inheritance, it will be necessary to replace the old values and habits of industrial expansion and affluence and create a more rational, frugal but also more equitable world in the twenty-first century.

▼ Notes

1. In the United States, for example, military expenditures are 6.6 percent of GNP. Japan spends 1.0 percent of GNP on the military, West Germany 3.1 percent, and the United Kingdom 5.1 percent. Figures are from the Center for Defense Information and published in Gerald Seib, "Candidates' Calls for Allies to Share Defense Costs Raises Complex Questions about U.S.'s World Role," *The Wall Street Journal* (March 2, 1988).

2. See Deborah Olivier, "Few Industries Benefit from the Weaker Dollar," *The Wall Street Journal* (January 30, 1987).

3. Tim Carrington and Robert Greenberger, "Pentagon's Firm Control Over Export Licenses May Lessen in Face of Politics, Big Trade Gaps," *The Wall Street Journal* (March 13, 1987).

4. Steven Greenhouse, "The Average Guy Takes It on the Chin," *The New York Times* (July 13, 1986). Similar figures show that inflation-adjusted average weekly earnings in private nonagricultural industries peaked at $198 per week in 1973 and steadily eroded to $169 weekly by 1987. See *Economic Report of the President 1988* (Washington, D.C.: U.S. Government Printing Office, 1988), p. 299.

5. John Culbertson, "Destructive Foreign Trade: Sowing the Seeds for Our Own Downfall," *The Futurist* (November–December, 1986), p. 13.

6. Ibid., p. 17.

7. Figures taken from "The Hollow Corporation," *Business Week* (March 3, 1986).

8. See Stephen Cohen and John Zysman, *Manufacturing Matters: The Myth of the Post-Industrial Economy* (New York: Basic Books, 1987), chaps. 1–2.

9. See Office of Technology Assessment, *International Competition in Services* (Washington, D.C.: U.S. Government Printing Office, 1987), pp. 57–69.

10. Percentages are derived from *Economic Report of the President 1987* (Washington, D.C.: U.S. Government Printing Office, 1987), table B-74.

11. See Paul Blustein, "Social Security Surpluses Said to Mask Size of Deficit," *The Washington Post* (March 3, 1988).

12. Figures are from Judith Havemann, "Uncle Sam's Math: 2 + 2 = 5," *The Washington Post* (April 19, 1987).

13. Lindley Clark, Jr., and Alfred Malabre, Jr., "Deep in Hock: Debt Keeps Growing With the Major Risk in the Private Sector," *The Wall Street Journal* (February 2, 1987).

14. Commerce Department data cited in Alfred Malabre, Jr., "Dependent Nation: U.S. Economy Grows Ever More Vulnerable to Foreign Influences," *The Wall Street Journal* (October 27, 1986).

15. *Economic Report of the President 1988*, p. 369.

16. Estimate made by C. Fred Bergsten and reported in Stuart Auerbach, "U.S. Becomes No. 1 Debtor Nation," *The Washington Post* (June 25, 1986).

17. Figure reported in Gene Koretz, "America's Dangerous Addiction to Capital from Abroad," *Business Week* (December 30, 1985).

18. Figures from Edward Foldessy and Tom Herman, "Analysts Fret over Holdings of Debt Issues by Foreigners," *The Wall Street Journal* (February 24, 1986).

19. John Burgess, "As Foreign Investment Increases, So Do Concerns About Its Impact," *The Washington Post* (February 21, 1988).

20. See Martin Tolchin and Susan Tolchin, *Buying into America* (New York: Times Books, 1988), pp. 268–274.

21. See Jethro K. Lieberman, *The Litigious Society* (New York: Basic Books, 1981).

22. See "The Death of Mining: America Is Losing One of Its Basic Industries," *Business Week* (December 17, 1984).

23. U.S. Patent Office data reported in Malcolm Gladwell, "Foreigners Get 46.6% of U.S. Patents," *The Washington Post* (February 26, 1988).

24. Tim Carrington and Robert Greenberger, op. cit.
25. National Science Board, *Science Indicators: The 1985 Report* (Washington, D.C.: National Science Foundation, 1985), p. 6.
26. The problem of education and U.S. industry is discussed in Barbara Vobejda, "The New Cutting Edge in Factories: Education," *The Washington Post* (April 14, 1987).
27. See the macroeconomic analysis of Lester Thurow, *The Zero Sum Solution* (New York: Simon & Schuster, 1985), chap. 10.
28. One of the strongest arguments from this perspective has been made by John Culbertson, op. cit.
29. These arguments are cogently made in John Zysman and Laura Tyson, *American Industry in International Competition* (Ithaca, N.Y.: Cornell University Press, 1983), chap. 1.
30. See Lester Brown, "A Generation of Deficits," in Lester Brown, ed., *State of the World 1986* (Washington, D.C.: Worldwatch Institute, 1986).
31. Richard Sterling, *Macropolitics* (New York: Knopf, 1974), p. 6.

▼ Suggested Reading

YAIR AHARONI, *The No-Risk Society* (Chatham, N.J.: Chatham House, 1981).

JAMES BOTKIN ET AL., *Global Stakes: The Future of High Technology in America* (Cambridge, Mass.: Ballinger, 1982).

STEPHEN COHEN AND JOHN ZYSMAN, *Manufacturing Matters: The Myth of the Post-Industrial Economy* (New York: Basic Books, 1987).

MICHAEL CROZIER, *The Trouble with America: Why the System Is Breaking Down* (Berkeley: University of California Press, 1986).

AMITAI ETZIONI, *The Moral Dimension: Toward a New Economics* (New York: The Free Press, 1988).

MEL GURTOV, *Global Politics in the Human Interest* (Boulder, Colo.: Lynne Rienner, 1988).

CHALMERS JOHNSON, *The Industrial Policy Debate* (San Francisco: Institute for Contemporary Studies Press, 1984).

ROBERT LAWRENCE, *Can America Compete?* (Washington, D.C.: The Brookings Institute, 1984).

JETHRO LIEBERMAN, *The Litigious Society* (New York: Basic Books, 1981).

OFFICE OF TECHNOLOGY ASSESSMENT, *International Competition in Services* (Washington, D.C.: U.S. Government Printing Office, 1987)

MICHAEL PIORE AND CHARLES SABEL, *The Second Industrial Divide: Possibilities for Prosperity* (New York: Basic Books, 1984).

CLYDE PRESTOWITZ, JR., *Trading Places: How We Allowed Japan to Take the Lead* (New York: Basic Books, 1988).

ROBERT REICH, *The Next American Frontier* (New York: Times Books, 1983).

RICHARD STERLING, *Macropolitics* (New York: Knopf, 1974).

LESTER THUROW, *The Zero Sum Solution* (New York: Simon & Schuster, 1985).

MARTIN TOLCHIN AND SUSAN TOLCHIN, *Buying into America* (New York: Times Books, 1988).

JOHN ZYSMAN AND LAURA TYSON, *American Industry in International Competition* (Ithaca, N.Y.: Cornell University Press, 1983).

Index

TO THE OWNER OF THIS BOOK:

We hope that you have found *Global Technopolitics: The International Politics of Technology and Resources* useful.
So that this book can be improved in a future edition,
would you take the time to complete this sheet and return it? Thank you.

Instructor's name:

Department:

School and address:

1. What I like most about this book is:

2. What I like least about this book is:

3. My general reaction to this book is:

4. The name of the course in which I used this book is:

5. Were all of the chapters of the book assigned for you to read?

    If not, which ones weren't?

6. On a separate sheet of paper, please write specific suggestions for improving this book and anything else you'd care to share about your experience in using the book.

Optional:

Your name:_____ Date:_____

Your signature:_____

May Brooks/Cole quote you, either in promotion for *Global Technopolitics: The
International Politics of Technology and Resources* or in future publishing ventures?

   Yes: No:

   Sincerely,

   Dennis Pirages

NO POSTAGE
NECESSARY
IF MAILED
IN THE
UNITED STATES

BUSINESS REPLY MAIL
FIRST CLASS   PERMIT NO. 358   PACIFIC GROVE, CA

POSTAGE WILL BE PAID BY ADDRESSEE

ATT: *Dennis Pirages*

**Brooks/Cole Publishing Company
511 Forest Lodge Road
Pacific Grove, California 93950-9968**